Walberto Alberca Meza

Mango postharvest handling

Walberto Alberca Meza

Mango postharvest handling

in Tambogrande

ScienciaScripts

Imprint
Any brand names and product names mentioned in this book are subject to trademark, brand or patent protection and are trademarks or registered trademarks of their respective holders. The use of brand names, product names, common names, trade names, product descriptions etc. even without a particular marking in this work is in no way to be construed to mean that such names may be regarded as unrestricted in respect of trademark and brand protection legislation and could thus be used by anyone.

Cover image: www.ingimage.com

This book is a translation from the original published under ISBN 978-620-0-02467-1.

Publisher:
Sciencia Scripts
is a trademark of
Dodo Books Indian Ocean Ltd. and OmniScriptum S.R.L publishing group

120 High Road, East Finchley, London, N2 9ED, United Kingdom
Str. Armeneasca 28/1, office 1, Chisinau MD-2012, Republic of Moldova, Europe
Printed at: see last page
ISBN: 978-620-6-99511-1

GENERAL INDEX

DEDICATION

To the memory of my father Nicanor Bari, today converted into a Light of Life; an example of constant struggle, of tireless effort and of the best teachings.

To my beloved Mother Paula Abadita;

With gratitude and infinite love
for giving me life, a lot of patience,
unconditional support and immense love.

To my siblings: Joel, Yojany, Gilmer, Felimer, Segundo, Elvia and Elcy with whom I share the satisfaction of having completed this project.

THANK YOU

Carlos San Martin Zapata for his confidence as advisor of this work and the opportunity to guide me during the development of this project.

To the company BIOFRUIT SA for opening its doors to me and allowing me to develop this research, and I would especially like to thank the quality assurance area directed by my colleague Moisés Chapa R. for his support in each of the stages of the development of this thesis.

To the faculty of this prestigious Faculty of Agronomy, who were my trainers and thanks to their scientific contributions and admirable knowledge are now participants of my great dream of improvement.

To my parents, grandparents, siblings, as well as my good friends, classmates, workmates, and all those who in one way or another have contributed to the completion of this thesis.

SUMMARY

The research was conducted in the company BIOFRUIT SA located in the district of Tambogrande, Km 1076.1; Panamericana Norte. Carretera Tambogrande -Las Lomas, department of Piura, during the months of January to March 2016. The objectives were: To evaluate the effect of waxing and caliber on the quality and shelf life of Kent variety mango fruit under hydrothermal treatment for 90 minutes and to determine the effect of waxing and caliber factors.

The waxing factor was studied with the following levels: Ecowax export MG wax, Citrashine EU 3 wax, Decco Lustr 631 wax and no wax; the calibers evaluated were: Caliber 6 (646-700g), Caliber 7 (551-645g), Caliber 8 (481-550g) and Caliber 9 (426-480g). The Randomized Complete Block Design (BCA) was applied, in a 4x4 factorial arrangement with four replications, studying 16 treatments in total where the experimental unit consisted of a bunch of 15 mango fruits, and the determinations of fruit weight (g.), brix degrees, percentage of acidity, vitamin C and firmness or hardness of the pulp were carried out.

The conclusions reached were: The application of the different types of wax contributed to maintaining the quality of the fruit because the weight loss was lower when the DECCO LUSTR 631 wax was applied (decreased 4.34% at 42 days after harvest), followed by the treatment with CITRASHINE EU 3 (4.45%), compared to 7.95% for the control treatment (WITHOUT WAX). Fruit firmness was also higher, registering 8.06kgf for treatment E3 corresponding to DECCO LUSTR 631 wax, compared to 5.94kgf for the treatment without wax. The total soluble solids measured in brix degrees, the percentage of acidity and the vitamin C content were also evaluated, where it was found that the size and waxing factors, in most of the evaluations, had no significant effect or their influence was minimal, being more difficult the action of waxes on the chemical constitution of the fruit of the Kent variety mango. In the size factor, it was found that weight loss is directly related to fruit size and weight, i.e. larger and heavier fruits (size 6) lose more weight (34.4 g) and vice versa (size 9) lost a total of 26 g, which in percentage represents 5.12% and 5.69% respectively. The interaction Caliber by Waxing was not statistically significant, so it is concluded that the factors act independently.

Key words: size, waxing, mango fruit, hydrothermal treatment.

INTRODUCTION

Mango is a climacteric fruit, which continues its metabolic processes even after being harvested; where enzymes and storage temperature influence the shelf life of this fruit. The use of waxy coatings which are available in the market for packers as mentioned by Mendoza (2005) decrease the respiration rate which is an indicator of metabolic activity and this leads to maturity, therefore, the use of waxes and polymers will improve the storage and marketing conditions of this fruit.

Taking into account the weight losses due to dehydration and other nutritional losses, as well as the loss of appearance that fruits face between the moment of harvesting and final consumption, it was considered convenient to carry out this research using three types of wax, in order to determine qualitatively and quantitatively the behavior of some quality parameters such as weight, firmness, brix degrees, acidity and vitamin C, in fruits of different sizes or calibers subjected to hydrothermal treatment for 90 minutes.

Considering that in the last seventeen years, according to the (Ministry of Agriculture and Irrigation [MINAGRI], 2017) Peruvian mango exports show a very dynamic behavior, with annual rates of 12.6% to the world, 17.6% to the European Union, 22.3% to the Netherlands and 7.2% to the United States, making Peru the fourth largest exporter in the world market, displacing Brazil and expecting to equal the volume exported by Thailand and India by 2021; it is necessary to conduct research on the post-harvest handling of this product in order to obtain new alternatives to help improve competitiveness in the market, so the information obtained in this research will be an important contribution mainly for agroexporting companies that develop this activity in the Piura region as well as for professionals and technicians interested in the subject.

CHAPTER I

ASPECTS OF THE PROBLEM

1.1. DESCRIPTION OF THE REALITY OF THE PROBLEM

Different types of wax are currently sold in the market, which are used by the packing plants to preserve the fruit after harvest. The fruits are subjected to hydrothermal treatment with different immersion times according to their size and weight, called caliber, and the quality conditions in which the exported fruit arrives at its destination are unknown.

Likewise, there is no local information on the use of waxes on hydrothermally treated mango fruit and its relationship with the quality of the fruit. Having this information will be important for the companies that export this product since it will allow them to give technical support on the use of waxes as post-harvest treatment of mango fruit. Therefore, from the above mentioned, the research problem is: ^How and to what extent the types of wax applied on mango fruits of the Kent variety of different sizes influence the shelf life and quality of the fruits in post-harvest.

1.2. JUSTIFICATION AND IMPORTANCE OF RESEARCH

The cultivation of mango (***Mangifera indica*** L.) according to MINAGRI 2016; in Piura were registered 19,896 Has. Cultivated of mango of which 98% is of the Kent variety; being the variety on which the largest volume of exports is made because it has a quite acceptable market, by the resistance of its shell, pleasant flavor; aroma and low fiber content.

The fresh fruit packing companies currently use different types of waxes and to date there is no research work that quantifies the benefits of the use of waxes on the reduction of weight loss and other quality parameters of the fruit; likewise, fruit of different sizes or calibers are packed, about which the behavior of the different types of waxes is also unknown.

Knowledge of the weight losses of the fruit, before and after waxing, will help us to complement the study, being this work of vital importance for the people who are part of the mango agro-export chain, whether they are producers, stockpilers, exporters, students, professionals, etc.

1.3. OBJECTIVES

1.3.1. General Objective

To evaluate the effect of waxing and size of Kent variety mango fruit on shelf life and fruit quality under hydrothermal treatment for 90 minutes.

1.3.2. **Specific Objectives**

> To evaluate the effect of waxing on the quality and shelf life of Kent variety mango fruit under hydrothermal treatment for 90 minutes.

> To evaluate the effect of size on the quality and shelf life of Kent variety mango fruit under hydrothermal treatment for 90 minutes.

> To evaluate the effect of the interaction of the factors waxing and caliber on the quality and shelf life of mango fruit of the Kent variety under hydrothermal treatment for 90 minutes.

1.4. DELIMITATION OF THE RESEARCH

1.4.1. **Spatial delimitation:**

The present research work was carried out in the company BIOFRUIT SA located at Km 1076.1 Panamericana Norte. Tambogrande - Las Lomas Highway. Whose political location corresponds to the department of Piura, province of Piura, district of Tambogrande, San Lorenzo Valley.

1.4.2. **Time delimitation**

It began on January 18 with the harvesting of the fruit and culminated on March 1, 2016 with the removal of the fruit from the company's storage chamber.

1.4.3. **Geographical delimitation (UTM coordinates):**

Longitude : 573767; Latitude: 9457171; Altitude : 69m

CHAPTER II

THEORETICAL FRAMEWORK

2.1. RESEARCH BACKGROUND

ALARCON (1994), in his thesis work entitled "methods of conservation, waxing, bagging and refrigeration in mango post-harvest"; using Primafresh wax, concluded that the best response in terms of conservation time is obtained in waxed fruits and refrigerated at 10°C.

CACERES et al (n.d.) in their work entitled "Influence of waxing and heat treatment on postharvest quality of mango" of the Tropical Fruit Research Institute (Instituto de Investigaciones en Fruticultura Tropical). It concludes that the application of carnauba waxes and sucrose ester maintains quality, contributing to preserve firmness and reducing weight losses of mango fruits in postharvest.

CALLE (1999) concluded that vacuum packing with or without hydrothermal treatment resulted in a longer shelf life up to 42 days after harvest.

PUELLES (2006), in a study on the effect of conservation methods on mango fruits (*Mangifera indica* L.) cultivars Kent and Haden, applied in post-harvest"; used Premium wax diluted in water in a proportion of three parts wax to one part water and applied individually on Kent mango fruits with a sponge where he determined that the weight loss was greater for the unwaxed treatment with 17.28% compared to the waxed treatment (6.43%).

2.2. THEORETICAL BASIS

2.2.1. Taxonomy

SAMSON (1991), states that the mango has the following taxonomic classification:

Reino	Plantae : Plantae
Division	Phanerogamma: Phanerogamma
Sub division	: Angiospermae
Class	: Dicotyledonea
Order	: Sapindales
Family	Anacardiacea : Anacardiacea
Sub family	Anacardioideae
Genre	Mangifera
Species	*Mangifera indica* L. cv. Kent

2.2.2. Ecology

WOLFE, OORT AND OTHERS (1969), state that mango grows well in warm climates, in tropical and subtropical zones where the mean annual temperature ranges between 20°C and 25°C. They also mention that the crop grows in various types of soils, as long as they have good drainage. Plants should preferably be established in soils with a clay loam or sandy loam texture, with a water table depth of 1.5 to 2.0 m, and a pH of 5.5 to 7.5.

2.2.3. Origin

FRANCIOSI (1991), reports that mango is a tropical species native to Southwest Asia originating in the forests of the Himalayan Mountains of India and the western part where it spread to Vietnam, Indonesia, Ceylon, Pakistan.

2.2.4. Quality of harvested fruit

ATARAMA (2008), evaluated the number of sun-stressed fruits per treatment, with the following scale: light red, yellowish and whitish, expressed in cm^2 the affected area.

- Grade 0: no insolation

- Grade 1 : onset of insolation only around the lenticels.

- Grade 2: insolation with union of the surroundings of the lenticels.

- Grade3 :< $1cm^2$

- Grade4 :1-2cm^2

- Grade5 :> $2cm^2$

Proposed scale for measuring insolation, in degrees:

Grade 0 Grade 1Grade   2

2.2.5. Characteristics of the fruit to be discarded in the first line in the plant for fresh mango exports

SAN MARTIN 2012, indicates that the discarding of fruit in the first line is carried out taking into account the following defects:

- With black necrotic spots or streaks

- With queresas, fruit with physiological pores

- Fruit without purple plate (less than 30%)

- With latex stains

- With sunstroke or sunburn

- With recent or old bruises or cracks

- With very small peduncle

- Fruit with the appearance of having been harvested days before (very dry stalk).

- Very small (<350 g.) or very large fruit (greater than 650 g.)

- Green fruit, round mangoes.

- Fruit with harvest or transport shock

- Fruit with signs of anthracnose and/or oidium

- Pale red Tommy Atkins fruit

- Fruit with soft consistency.

- Fruit with calcium deficiency symptoms

- Mangoes with suspected fruit fly.

2.2.6. Sizing table

In Annex 37, AGROMAR INDUSTRIAL (2015), shows the sizing table used for mango for export. Likewise, BIOFRUIT (2015), indicates that sizing consists of separating the fruits according to their size and weight, which indicate the number of units to be packed in a 4 kg box.

BIOFRUIT also states that from the 2015-2016 campaign, it will be possible to work with the new table of times and weights for hydrothermal treatment of mangoes approved by USDA-APHIS.

On 04-08-2015. The United States Department of Agriculture (USDA) published in the Federal Register (FEDERAL REGISTER) the order of immediate modification of the work plans of the mango exporting countries to the United States, incorporating the new table of times and weights to be used shown below:

Dive time (Minutes)	Weight range (g.)
65	up to 375gr.
75	from 375 to 500 gr
90	from 501 to 700gr
110	from 701 to 900gr.

Source: APEM News.

2.2.7. Waxing treatment on mango fruit

ACOSTA(1988), indicates that the use of natural or synthesis waxes is also a post-harvest treatment aimed at extending the commercial life of the fruits, protecting them against fungal rots, reducing water losses by transpiration and giving them a more attractive appearance to the consumer, experiences have shown a significant extension in the storage period of mangoes, which were treated with a 6% wax emulsion and stored at room temperature.

MENDOZA(2005), mentions that there are products on the market available to packers that reduce the respiration rate, such as waxes and polymers that, if used, will improve mango storage and marketing conditions.

YAHIA(1992), states that waxing could be used as a preservation method for citrus and other products.

2.2.8. Hydrothermal treatment

FRANCIOSI (1992), points out that in the hot water or hydrothermal treatment, the fruit is immersed in specially prepared tanks containing water maintained at a temperature of 46°C; the immersion time varies between 70 and 90 minutes depending on fruit size and harvest time. Control is effective on larvae of *Ceratitis capitata, Anastrepha fraterculus, Anastrepha oblicua and Anastrepha disticta.*

MENDOZA (2005), states that the treatment of fresh mangoes in hot water at 115°F (46.1°C), is a quarantine process aimed at eliminating the different imperfect larval stages of the fruit fly (*Anastrepha sp*), the method consists of subjecting fresh mangoes to a pre-determined temperature (set point), through immersion in hot water of continuous and automatic flow in tanks or tubs for this purpose.

2.2.9. Shelf life

PUELLES (2006) recommends using the treatment of waxed and refrigerated cv. kent fruit to maintain a relatively constant weight and conserve fruit quality. He also suggests carrying out trials with different doses of wax and different sizes.

Shelf life is the time required for a product, under certain storage and packaging conditions, to deteriorate to an unacceptable state or to become unsuitable for marketing.

2.3. HYPOTHESIS

2.3.1. General hypothesis

Waxing and caliber influence the quality of Var. Kent mango fruit treated by hydrothermal treatment for 90 minutes.

2.3.2. Specific hypotheses

•	Waxing influences the quality and shelf life of Kent variety mango fruit under hydrothermal treatment for 90 minutes.

•	Size influences the quality and shelf life of Kent variety mango fruit under hydrothermal treatment for 90 minutes.

•	The joint action of waxing and sizing factors affect the quality and shelf life of Kent variety mango fruit under hydrothermal treatment for 90 minutes.

CHAPTER III

METHODOLOGICAL FRAMEWORK

3.1. APPROACH AND DESIGN

The present research has a quantitative approach of experimental type. Quantitative because it is made of variables with response to the application of treatments. Experimental, because the evaluation of the treatments was done using an experimental design, a model and statistical tests that help to explain the behavior of the treatments under study.

3.2. RESEARCH SUBJECTS

3.2.1. Experimental unit

The experimental unit was represented by a bale containing 15 mango fruits. Three fruits were marked on each bunch, these were weighed every week for seven weeks, and one mango fruit was taken at random each week. Taking into account that there were four treatments (including the control), four sizes and four replications, a total of 64 experimental units of 15 fruits each were made, that is, 960 fruits, which were fully identified for their respective evaluations.

3.3. METHODS AND PROCEDURE

3.3.1. Materials and equipment

a. Materials

- Mango fruit of the Kent cultivar.

- Plastic beads.

- Calculator, notepad, pencil, pencil

- Knife, cutting board

- Distilled water, blotting paper, industrial wiping cloth

- Hydraulic rope, hydraulic rope, straps

- Calcium hypochlorite

- Thiabendazole

b. Equipment

- Precision balance, brixmeter, penetrometer

- Hydrothermal treatment equipment

- Cooling tunnel, cold storage chamber, extractor liquefuge

- Digital camera, computer, complete laboratory equipment.

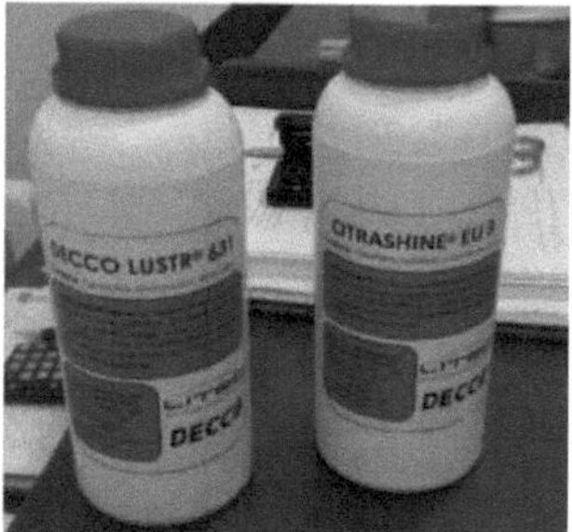

Figure 3.1. Packaging of DECCO LUSTR 631 and CITRASHINE EU 3 waxes.

3.3.2. Procedures

3.3.2.1. Visit and inspection of Kent mango plantation

A visit was made to the plot belonging to the producer Efrain Agurto Ruiz, who receives technical assistance from the company BIOFRUIT SA, located in the Hualtaco III sector in front of the Pan-American Highway North Tambogrande-Las Lomas road of approximately six hectares of Kent mango certified for export in order to determine physical indices of maturity of the fruit and schedule the harvest date.

3.3.2.2. Harvest

The harvest was carried out manually using rustic ladders, cutting shears and plastic scissors to collect the fruit.

Figure.3.2. Hand-harvested material

Figure 3.3. Harvesting

3.3.2.3. Scrapping

The fruit was cut down to a size of approximately 0.5 cm, as required by export quality parameters, and then placed with the stalk cut downwards on a guayaquil grid so that the latex exuding from the cut fruit does not touch the skin. The fruits were then placed with the stalk cut downwards on a guayaquil grid so that the latex exuded by the cut fruit does not touch the peel and there is no direct contact with the ground.

Figure 3.4. Fruits after stalk cutting.

3.3.2.4. Borax application

Approximately one hour after scrapping, a chemical product commonly known as borax was applied to the stalk, which allows the stalk to stop exuding latex and thus reduce the damage caused by this substance to the fruit shell.

Figure 3.5. Application of borax

3.3.2.5. Transport of the fruit to the packing house

Thirty jabas of Kent mango were separated and properly stowed in a truck that was in charge of transporting the fruit to the BIOFRUIT SA plant. Where the research was carried out.

Figure 3.6. Transport of fruit to packing plant

3.3.2.6. Washing, disinfection and fruit sizing

This process was carried out using the washing tub which was dosed with calcium hypochlorite at a concentration of 80 ppm, continuing the process by brushing and a sprayer containing the fungicide thiabendazole at a dose of 1ml/liter of water.

Sizing was carried out by separating the fruits according to their size and weight according to the weight chart used by BIOFRUIT SA. during the 2015-2016 campaign. Calibers 6, 7, 8 and 9 were separated, which are the ones used in the research.

Figure. 3.7. Beginning of fruit washing process

3.3.2.7. Hydrothermal treatment

With the samples fully identified, the fruit was immersed in hot water in the hydrothermal treatment area of BIOFRUIT SA, certified by SENASA and APHIS for export of fresh mango to USA, which lasted 90 minutes as required by the plant quarantine protocol for exporting fresh mango to USA. After removing the fruit from the hydrothermal treatment was allowed to cool for a period of six hours to then apply the waxing.

Figure.3.8. Fruit to be subjected to hydrothermal treatment

3.3.2.8. Application of waxing

Once the fruit had been cooled for 6 hours, the experimental units corresponding to the control treatment were separated and placed in a parihuela, to which no wax was applied, leaving the fruit ready to be taken to the storage chamber for conservation, where it was kept at an average temperature of 9°C.

The three types of wax used according to the manufacturer's indications are coatings that come prepared and ready to be applied without the need to be diluted or mixed with other products; therefore, 100% wax was added in a tank which was adapted to the mechanical and electrical waxing machine used by BIOFRUIT SA. Then the fruit was introduced into the vat, which finally, by means of rotating brush rollers, made the fruit advance homogeneously and was passed through the sprinkler in charge of applying the wax that was added.

The application began with the tank containing the ECOWAX EXPORT MG wax; when the waxing was finished, all the fruit corresponding to this treatment was collected in jabas and placed in the same parihuela in which the control was placed to be taken to the storage chamber. This was followed by the CITRASHINE EU3 wax following the same procedure and finally the DECCO LUSTR 631 wax treatment was applied in the same way.

Figura 3.9. Fruit passing through the waxing machine

3.9.2.9. Transfer of treatments to the storage chamber

All fully identified treatments were placed in a single pair and finally taken to the storage chamber at 9°C where they remained for their respective evaluations until the end of the experiment.

Figura 3.10.   Fruit in storage chamber at 9°C temperature

3.3.3. Factors under study

The factors and levels under study are presented in Table 3.1.

Table 3.1. Factors and levels under study

FACTOR	LEVEL	KEY
Waxing	Without waxing	E $_0$
	Ecowax Export MG	E $_1$
	Citrashine EU 3 Wax	E $_2$
	Lustr Wax 631	E $_3$
Caliber	6 (646-700 g.)	C $_1$
	7 (551-645g.)	C $_2$
	8 (481-550 g.)	C $_3$
	9 (426-480g.)	C $_4$

Figure 3.11. 6-gauge fruits

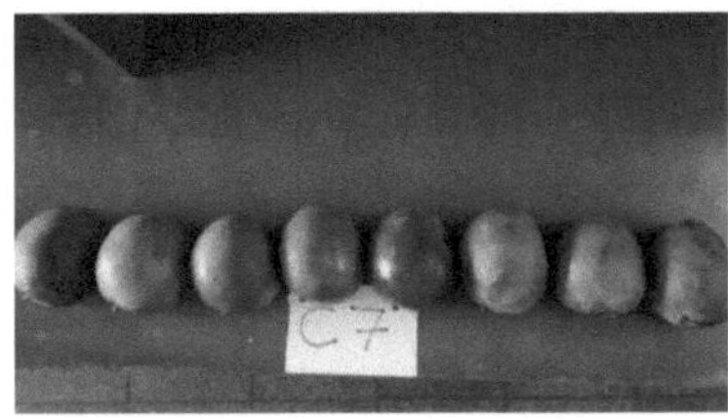

Figuran 3.12. 7-gauge fruit

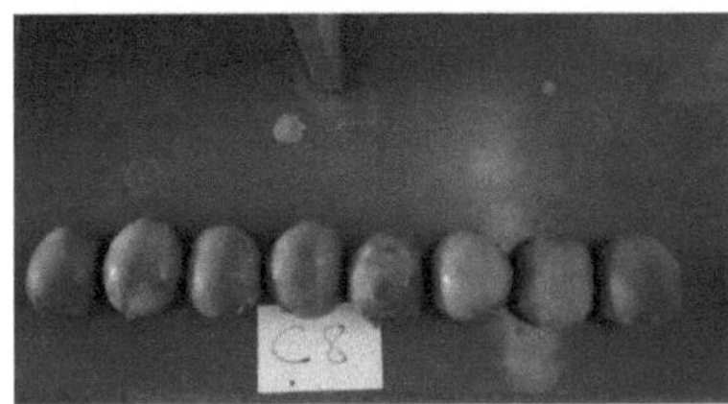

Figure 3.13. 8-gauge fruit

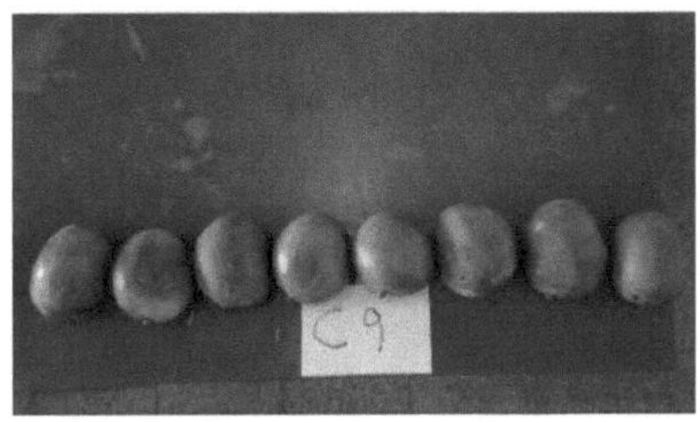

Figure 3.14. Fruits caliber 9

3.3.4. Treatments under study

The treatments under study are shown in Table 3.2.

Table 3.2. Treatments under study

N°	DESCRIPTION	KEY
1	No waxing on 6-gauge fruit	$E_0 C_1$
2	No waxing on 7-gauge fruit	$E_0 C_2$
3	No waxing on 8-gauge fruit	$E_0 C_3$
4	No waxing on 9-gauge fruit	$E_0 C_4$
5	Wax Ecowax Export MG in 6-gauge fruit	$E_1 C_1$
6	Wax Ecowax Export MG in 7-gauge fruit	$E_1 C_2$
7	Wax Ecowax Export MG in 8-gauge fruit	$E_1 C_3$
8	Ecowax Export MG wax in 9 gauge fruit	$E_1 C_4$

9	Citrashine wax in 6-gauge fruit	$E_2 C_1$
10	Citrashine wax in 7-gauge fruit	$E_2 C_2$
11	Citrashine wax in 8-gauge fruit	$E_2 C_3$
12	Citrashine wax in 9 gauge fruit	$E_2 C_4$
13	Decco Lustr 631 wax in 6-gauge fruit	$E_3 C_1$
14	Decco Lustr 631 wax in 7-gauge fruit	$E_3 C_2$
15	Decco Lustr 631 wax in 8-gauge fruit	$E_3 C_3$
16	Decco Lustr 631 wax in 9-gauge fruit	$E_3 C_4$

3.3.5. Experimental design

The Randomized Complete Block design (RCB) was used, in a 4x4 factorial arrangement with four replications. A total of 16 treatments were studied. Each experimental unit consisted of a bunch of 15 mango fruits.

Figure 3.15. Arrangement of treatments

3.3.6. Experimental determinations

The determinations were carried out at the Laboratory of Analysis of Agricultural Products of the Department of Agronomy and Phytotechnology of the National University of Piura.

The evaluations were carried out from the beginning of the application of the treatments, with a frequency of 7 days until completing 40 days after waxing.

Figure 3.16. Samples for determinations

1. Fruit weight (g.)

The fruits were weighed using a digital balance, as shown in Figure 3.17.

20

Figure 3.17. Evaluation of initial fruit weight.

2. Brix degrees (°Brix)

For this determination, the fruits were cut in order to extract two drops from the pulp.

of juice and place them in the refractometer and take °Brix content readings.

Figure 3.18. Taking °brix readings

3. Percentage of acidity

It was determined by titration with 0.1N sodium hydroxide, for which the mango juice had to be extracted in a liquefying extractor, taking 10 ml of sample to carry out the acidity determination procedure.

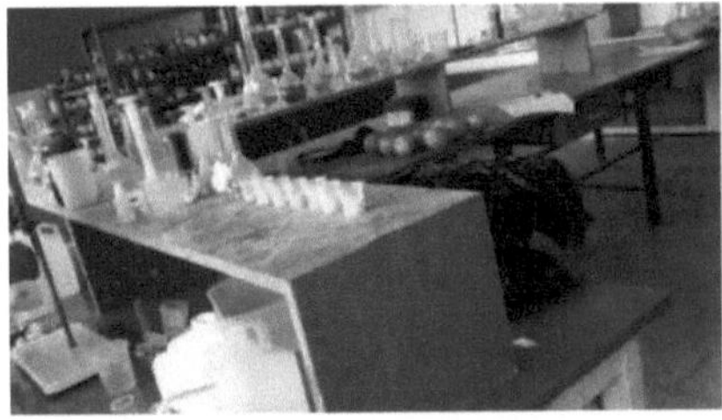

Figure 3.19. Determining acidity

4. Vitamin C (mg ascorbic acid/100cc of mango juice)

It was determined by titration using 0.1N iodine; using a sample of juice from the fruit extracted with an iodine extractor.

Figure 3.20. Determining vitamin C

5. Firmness or hardness of pulp (Kgf)

It was determined with the penetrometer by direct measurement.

Figure 3.21. Determining fruit firmness

3.4. TECHNIQUES AND INSTRUMENTS

3.4.1. Techniques

a. Sampling

The sampling technique used was simple sampling, where within the experimental unit the elements of the population had the same possibility of being taken for evaluation. In the evaluations, mango fruits were taken at random.

b. Data collection

The data collection of the elements of each experimental unit was done by direct measurement. The units of measurement used were the gram, brix degrees, acidity percentage and mg. of Ascorbic Acid/100 ml of juice for vitamin C content and Kgf for firmness.

3.4.2. Instruments

Weighing was carried out using precision scales with a maximum tare weight of 5.0 kg. Brix and firmness data were obtained using a refractometer and penetrometer, respectively. The determination of acidity and vitamin C was carried out using material and equipment from the seed laboratory of the academic department of Agronomy and Phytotechnology of the National University of Piura.

3.5. ETHICAL ASPECTS

The present work was carried out respecting the norms of good manufacturing practices, standards of behavior, hygiene and sanitation; in addition to using the corresponding protective clothing and equipment in order to ensure food safety and quality policies implemented by the company BIOFRUIT SA.

CHAPTER IV

RESULTS AND DISCUSSION

4.1. FRUIT WEIGHT (g.)

Tables 4.1 a and 4.1 b show the results of the mean squares and statistical significance in eight evaluations of fruit weight of the Kent mango variety, studied with four sizes and four types of waxing; it can be seen that in all of them, statistical significance was obtained at the 0.01 probability level in the size factor, while in the waxing factor, there was a significant response in only three evaluations, one of them at the 0.05 level and the rest at the 0.01 level.

No significance was detected in the C x E interaction, which would indicate that the two factors studied were acting independently on this characteristic.

Variation coefficients ranged from 1.68% to 2.81%, values considered quite low, so the information reported is reliable. The records for this characteristic are shown in Annexes 01 to 08.

The Duncan$0._{05}$ tests performed (Table 4.2a) confirmed the results of the respective ANVAS for the size factor, showing that size-6 reported the highest mango weights from the first to the eighth evaluation, statistically surpassing the weights obtained by the other sizes. Similarly, it can be seen that size-9 presented the lowest fruit weight records in all the evaluations carried out.

It is also possible to observe that as the evaluations were carried out, the weight of the mango fruit decreased, regardless of the size analyzed, with greater losses in size 6, and these losses diminished as the size increased. In addition, it was observed that weight losses were greater in the first two evaluations, and then continued, but to a lesser extent. On average, these losses were in the order of 5.42% (see Table 4.2 b).

When analyzing the data of fruit weight according to the type of waxing, it is observed that during the first three evaluations, no significant statistical differences were detected, being therefore the values of mango weights, quite similar within each evaluation, but are progressively decreasing between evaluations, with higher losses in the control treatment (E_0), which did not receive any type of wax (7.95%), because not having the mango fruits, a protector, the decrease in weight will be greater, than if they were protected.

In the other types of waxing, weight losses were lower, from E_i, when ECOWAX EXPORT MG was applied, which registered 4.81%, then E2 (CITRASHINE EU3), with a loss of 4.45%, and finally E3 (LUSTR 631), with the lowest value of weight loss (4.34%).These results confirm the indications of **ACOSTA(1988),** who states that the use of natural wax emulsions reduces water loss by transpiration, and **YAHIA(1992),** who indicates that waxes are a preservation method for citrus and other products, among which we have experimented with mango.

Again, the greatest weight losses occurred during the first two evaluations, and continued thereafter, but to a lesser extent (see Tables 4.3 a and 4.3 b).

For a better understanding of the above, see figures 4.1 to 4.4.

Table 4.1.a. Summary of mean squares and statistical significance in four evaluations of mango fruit weight Var. Kent (g.)

F. of Variation	G.L	EVALUATION			
		CM (1) 19 - 01 - 2016	CM (2) 20 - 01 - 2016	CM (3) 25 - 01 - 2016	CM (4) 01 - 02 - 2016
Blocks	3	21.62 ns	13.59 ns	16.32 ns	11.38 ns
Treatments	(15)				
Calibers	3	139302.10 **	135322.26 **	132933.24 **	131523.29 **
Waxing	3	50.79 ns	62.35 ns	150.10 ns	309.26 *
C x E	9	105.83 ns	121.15 ns	112.02 ns	105.74 ns
Experimental error	45	90.64	88.46	85.04	99.68
Total	63	CV (%) = 1.68	1.69	1.68	1.83

Note: ns = Not significant

* Statistical significance at 0.05 probability level.

** Statistical significance at 0.01 probability level.

Table 4.1.b. Summary of mean squares and statistical significance, in four evaluations of mango fruit weight Var. Kent (g.).

F. of Variation	G.L	EVALUATION			
		CM (5) 08 - 02 - 2016	CM (6) 15 - 02 - 2016	CM (7) 22 - 02 - 2016	CM (8) 29 - 02 - 2016
Blocks	3	102.29 ns	21.16 ns	208.36 ns	6.84 ns
Treatments	(15)				
Calibers	3	129937.81 **	130438.28 **	126550.65 **	127972.25 **
Waxing	3	234.47 ns	624.14 **	488.44 ns	1319.81 **
C x E	9	152.30 ns	128.89 ns	193.39 ns	80.98 ns
Experimental error	45	86.74	91.50	226.50	93.74
Total	63	1.72	1.78	2.81	1.82

Note: ns = Not significant

* Statistical significance at 0.05 probability level.

** Statistical significance at 0.01 probability level.

Table 4.2.a. Summary of Duncan0.05 tests for fruit weight of mango Var Kent, during eight evaluations, in four sizes (g.).

CALIBER	EVALUATION							
	1	2	3	4	5	6	7	8
Caliber - 6 (646 - 700)	672.3 a	662.7 a	655.6 a	651.2 a	647.4 a	644.4 a	640.2 a	637.9 a
Caliber - 7 (546 - 645)	600.2 b	591.1 b	584.3 b	580.0 b	576.3 b	574.3 b	568.9 b	567.1 b
Caliber - 8 (481 - 545)	524.2 c	516.2 c	510.9 c	507.3 c	504.4 c	501.4 c	497.7 c	496.1 c
Caliber - 9 (426 - 480)	456.8 d	450.3 d	444.7 d	441.4 d	438.8 d	435.6 d	432.2 d	430.8 d

Table 4.2.b. Var Kent mango fruit weight losses during eight evaluations, by size (g.).

CALIBER	EVALUATION							Total Loss	%
	1	2	3	4	5	6	7		
Caliber - 6	- 9.6	- 7.1	- 4.4	- 3.8	- 3.0	- 4.2	- 2.3	34.4	5.12
Caliber - 7	- 9.1	- 6.8	- 4.3	- 3.7	- 2.0	- 5.4	- 1.8	33.1	5.51
Caliber - 8	- 8.0	- 5.3	- 3.6	- 2.9	- 3.0	-3.7	- 7.8	28.1	5.36
Caliber - 9	- 6.5	- 5.6	- 3.3	- 2.6	- 3.2	- 3.4	- 1.4	26.0	5.69
AVERAGE	- 8.3	- 6.2	- 3.9	- 3.25	- 2.8	- 2.6	- 3.3	30.4	5.42

Note: Averages that have the same row are statistically equal, otherwise they are different.

Table 4.3.a Summary of Duncan's 0.05 tests of Var Kent mango fruit weight, during eight evaluations, in four waxing (g.)

WAXING	EVALUATION							
	1	2	3	4	5	6	7	8
E0	564.8 a	556.3 a	545.8 a	539.5 b	536.9 b	530.5 b	529.1 b	519.9 b
E1	560.8 a	552.4 a	546.7 a	543.4 ab	540.3 a	537.9 a	535.1 ab	533.8 a
E2	563.6 a	554.9 a	550.8 a	547.8 a	544.9 a	542.9 a	539.9 ab	538.5 a
E3	564.2 a	556.8 a	552.1 a	549.2 a	544.8 a	544.3 a	541.3 a	539.7 a

Note: Averages that have the same row are statistically equal, otherwise they are different.

Table 4.3.b. Var Kent mango fruit weight losses in eight evaluations by waxing **(g.).**

WAXING	EVALUATION		Total	%

	1	2	3	4	5	6	7	Loss	
E0	- 8.5	- 10.5	- 6.3	- 2.6	- 6.4	- 1.4	- 9.2	44.9	7.95
E1	- 8.4	- 5.7	- 3.3	- 3.1	- 2.4	- 2.8	- 1.3	27.0	4.81
E2	- 8.7	- 4.1	- 3.0	- 2.3	- 2.0	- 3.0	- 1.4	25.1	4.45
E3	- 7.4	- 4.7	- 2.9	- 4.4	- 0.5	- 2.0	- 1.6	24.5	4.34
AVERAGE	- 8.3	- 6.3	- 3.9	- 3.1	- 2.8	- 2.3	- 3.4	30.4	5.39

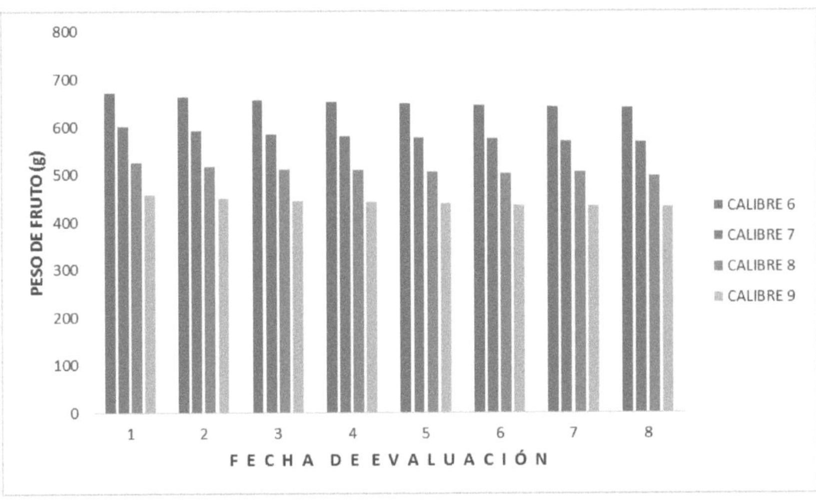

Figura 4.1.   Fruit weight of Kent mango variety of four sizes in eight evaluations (g.)

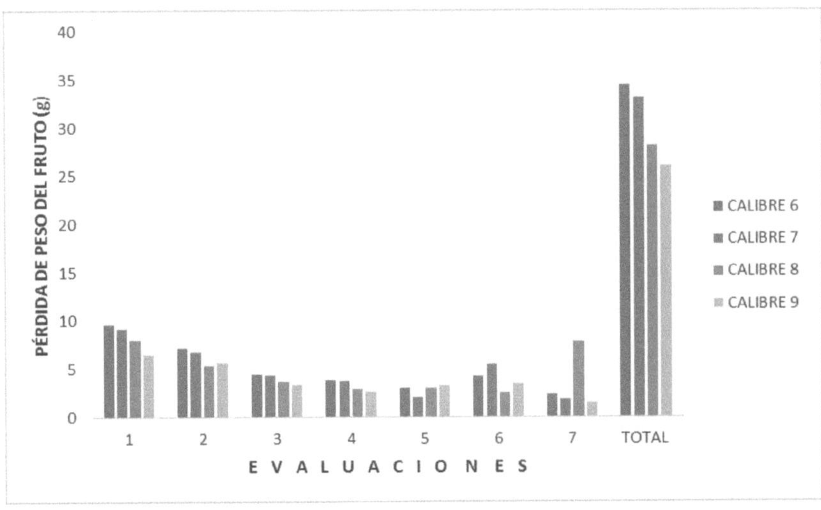

27

Figura 4.2.   Fruit weight loss of Kent variety mango fruit in four sizes in eight evaluations (g.)

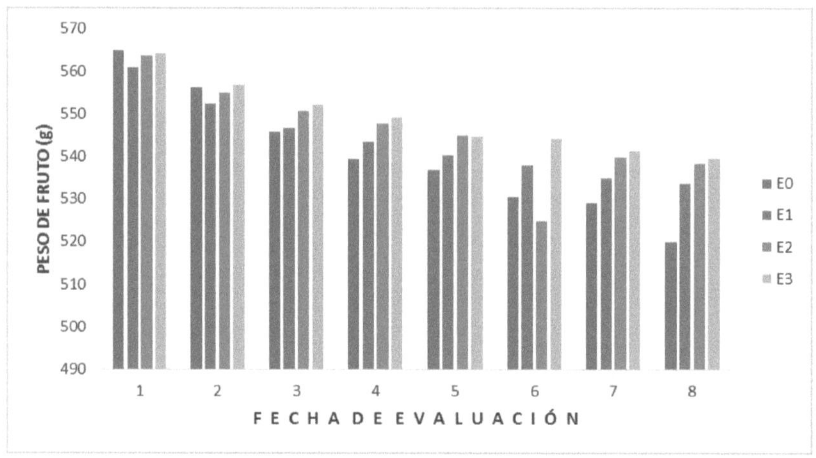

Figura 4.3.   Fruit weight of Kent variety mangoes from four waxings in eight evaluations (g.)

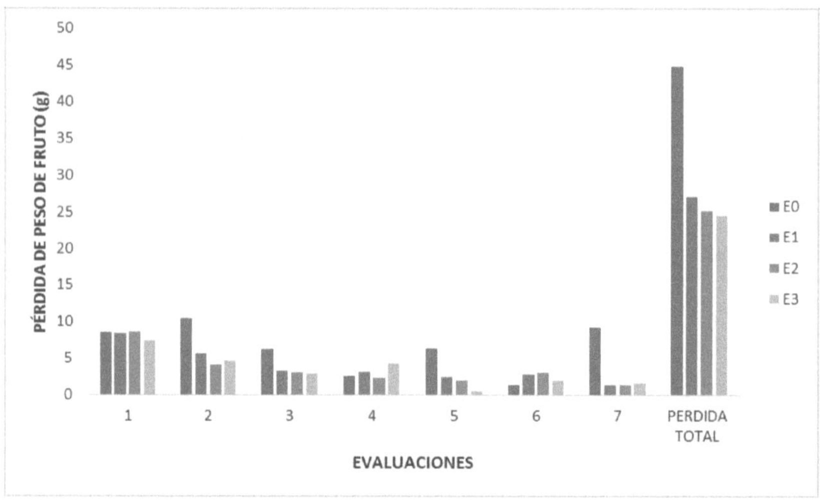

Figura 4.4.   Fruit weight loss of Kent variety mango fruit from four waxing in eight evaluations (g.)

4.2 DEGREES BRIX

The summary of the mean squares and their statistical significance, in seven evaluations carried out on the Kent mango variety, on the brix degrees (°), are shown in Tables 4.4 a and 4.4 b, from which the following can be deduced:

No statistical significance was detected in any of the sizes studied, while for the waxed varieties, all the evaluations showed significant statistical responses at the 0.01 level, with the sole exception of the first evaluation, which did not show a significant response.

The C x E interaction also did not respond significantly in any evaluation, which would indicate that both factors act independently on brix degrees (°).

The coefficients of variation (CV) ranged from 0.89% to 3.74%, which are considered low, therefore the information reported is valid. The original data for this characteristic can be found in Annex 9 through Annex 15.

Table 4.5 a, which shows the results of the DUNCAN$_{0.05}$ tests of the sieves studied on the brix degrees (°), practically confirms the results shown by the corresponding ANVAS.

Observing carefully the values of brix degrees (°), it can be seen that there is no tendency or behavior of the sizes, that is, there is no defined pattern of response, appreciating that the highest sugar contents are practically presented by the largest sizes (8 and 9), three each; and only in the first evaluation, the highest content of total sugars was presented by the smallest size (size-6= 6.95°), however, within each evaluation, the differences are practically only numerical (see Figure 4.5).

When studying the increase in brix degrees (°) of the four grades investigated during the seven evaluations carried out (Table 4.5 b), it is observed that the greatest increases occurred in the initial evaluations, and then decreased very sharply in the final evaluations, in some cases to less than 0.5°.

The highest gain of total sugars (8.92°) corresponded to caliber-9, representing 129.1% and together with caliber-8, its percentage gain was higher than the average value obtained (124.8%); while the percentage gain in the smaller caliber-6 and 7 was below the average (see Figure 4.6).

On the other hand, Table 4.6a shows the summary of the Duncan 0.05 tests performed to study the effect of waxes on brix degrees, confirming what was shown in Table 4.4a; that is, there are high statistical differences between waxes, with the exception of the Initial evaluation, where no statistical differences were reported.

Observing in detail the records shown, it can be seen that the Eo treatment (control or unwaxed) recorded the highest values of total sugars, due to the fact that the mango fruits were not protected, especially from evaluation three to the end.

Therefore, if we discard the control treatment (E0) and analyze only the values in the three types of waxes investigated, there is a tendency that the highest values of brix degrees are presented when

E1 (ECOWAX EXPORT MG) was used, and this would be an indicator that this product would be the least suitable for the preservation of mango fruits in the first evaluations; and in the last two evaluations, the best performance was again presented by the waxing-1 (ECOWAX EXPORT MG), which in this case reports the lowest data of brix degrees (°), and these would be the best values; and the worst performing waxing would be E2 (CITRASHINE EU3), since it is the one that presented the highest values of total sugars, which are the most undesirable. For a better understanding, see figure 4.7.

Finally, Table 4.6 b, shows the records of the increases in brix degrees of the Kent mango, by the waxed varieties, in the seven evaluations, showing the same behavior as that shown by the four sizes; that is to say, the greatest increases in total sugar occurred in the first two evaluations, and later these decreased very strongly, and there were only two cases, with the same Waxing (E1= ECOWAX EXPORT MG), that instead of increasing the brix degrees, the opposite happened, the values decreased, when passing from the fifth to the sixth and from this to the last evaluation.

The greatest increase in total sugars occurred in the control (Eo = without waxing), with an increase of 10.38°, due to its non-protection of the fruit, which represented 150.4%; while among the waxed varieties, the highest value (8.69°) was presented by E2 (CITRASHINE EU3), representing 124.7%. Figure 4.8 provides a better clarification of the above.

Table 4.4.a. Summary of mean squares and statistical significance, in four evaluations of brix degree in mango fruit Var. Kent.

F. of Variation	G.L	EVALUATION			
		CM (1) 18 - 01 - 2016	CM (2) 25 - 01 - 2016	CM (3) 01 - 02 - 2016	CM (4) 08 - 02 - 2016
Blocks	3	0.010 ns	0.295 ns	0.048 ns	0.013 ns
Treatments	(15)				
Calibers	3	0.006 ns	0.222 ns	0.079 ns	0.017 ns
Waxing	3	0.015 ns	2.141 **	4.596 **	4.175 **
C x E	9	0.027 ns	0.186 ns	0.025 ns	0.018 ns
Experimental error	45	0.029	0.194	0.048	0.016
Total	63	CV (%) = 2.46	3.68	1.64	0.89

ns = Not significant

* Statistical significance at 0.05 probability level.

** Statistical significance at the 0.01 probability level.

Table 4.4.b. Summary of mean squares and statistical significance, in three evaluations of brix degree in mango fruit Var. Kent.

F. of Variation	G.L	EVALUATION		
		CM (5) 15 - 02 - 2016	CM (6) 22 - 02 - 2016	CM (7) 29 - 02 - 201
Blocks	3	0.126 ns	1.303 **	0.953 ns
Treatments	(15)			
Calibers	3	0.115 ns	0.671 ns	0.942 ns
Waxing	3	5.058 **	9. 271**	26.494 **
C x E	9	0.107 ns	0.292 ns	0.539 ns
Experimental error	45	0.127	0.294	0.340
Total	63	2.40	3.54	3.74

ns = Not significant

* Statistical significance at 0.05 probability level.

** Statistical significance at the 0.01 probability level.

Table 4.5.a. Summary of Duncan $0._{05}$ tests of brix degree in mango fruit, Var Kent, during seven evaluations, in four sizes.

CALIBER	EVALUATION						
	1	2	3	4	5	6	7
Caliber - 6 (646 - 700)	6.95 a	11.80 a	13.36 ab	14.15 a	14.83 a	15.18 ab	15.27 a
Caliber - 7 (546 - 645)	6.91 a	11.96 a	13.28 a	14.15 a	14.71 a	15.16 a	15.49 ab
Caliber - 8 (481 - 545)	6.93 a	12.09 a	13.44 b	14.18 a	14.81 a	15.59 b	15.68 ab
Caliber - 9 (426 - 480)	6.91 a	11.98 a	13.30 ab	14.21 a	14.92 a	15.39 ab	15.83 b

Note: Averages that have the same row are statistically equal, otherwise they are different.

Table 4.5.b. Increase in brix degree in mango fruit, Var Kent, in seven evaluations carried out, by size.

CALIBER	EVALUATION						SUGAR PROFIT TOTAL	%
	1	2	3	4	5	6		-
Caliber - 6	+ 4.85	+ 1.56	+ 0.79	+ 0.68	+ 0.35	+ 0.09	8.32	119.7
Caliber - 7	+ 5.05	+ 1.32	+ 0.87	+ 0.56	+ 0.45	+ 0.33	8.58	124.2
Caliber - 8	+ 5.16	+ 1.35	+ 0.74	+ 0.63	+ 0.78	+ 0.09	8.75	126.3
Caliber - 9	+ 5.07	+ 1.32	+ 0.91	+ 0.71	+ 0.47	+ 0.44	8.92	129.1

AVERAGE	+ 5.03	+ 1.39	+ 0.83	+ 0.65	+ 0.51	+ 0.24	8.64	124.8

Table 4.6.a. Summary of Duncan 0.05 tests of brix degree in mango fruit, Var Kent, during seven evaluations, by waxing.

WAXING	EVALUATION						
	1	2	3	4	5	6	7
E0	6.90 a	11.91 a	13.82 b	14.75 d	15.49 a	16.36 c	17.28 d
E1	6.93 a	12.48 b	13.80 b	14.45 c	14.95 b	14.62 b	14.21 a
E2	6.97 a	11.87 a	12.87 a	13.66 a	14.13 c	15.43 a	15.66 c
E3	6.91 a	11.61 a	12.89 a	13.83 b	14.71 b	14.92 b	15.11 b

Note: Averages that have the same row are statistically equal, otherwise they are different.

Table 4.6.b. Increase in brix degree in mango fruit, Var Kent, in seven evaluations carried out by waxing.

WAXING	EVALUATION						Total increase	%
	1	2	3	4	5	6		
E0	+ 5.01	+ 1.91	+ 0.93	+ 0.74	+ 0.87	+ 0.92	+ 10.38	150.4
E1	+ 5.55	+ 1.32	+ 0.65	+ 0.50	- 0.33	- 0.41	+ 7.28	105.1
E2	+ 4.90	+ 1.00	+ 0.79	+ 0.47	+ 1.30	+ 0.23	+ 8.69	124.7
E3	+ 4.70	+ 1.28	+ 0.94	+ 0.88	+ 0.21	+ 0.19	+ 8.20	118.7
AVERAGE	5.04	1.38	0.83	0.65	0.51	0.23		

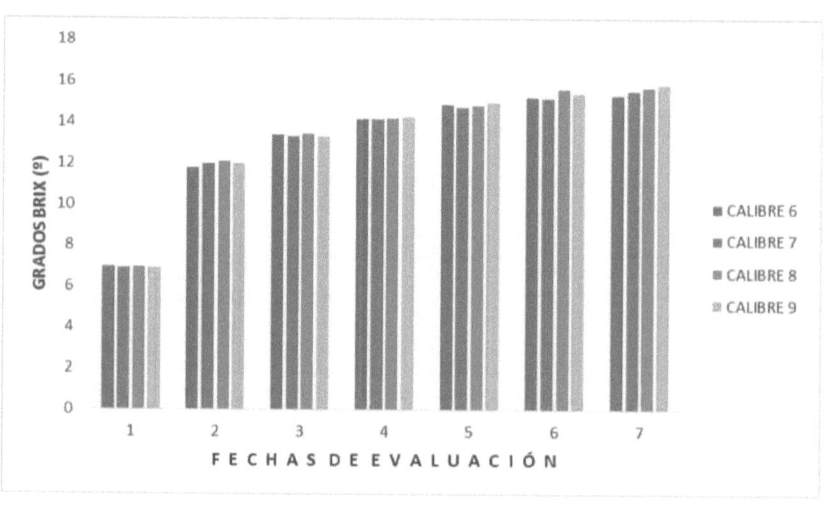

Figura 4.5. Brix degrees in mango fruits of Kent variety of four sizes in seven evaluations performed

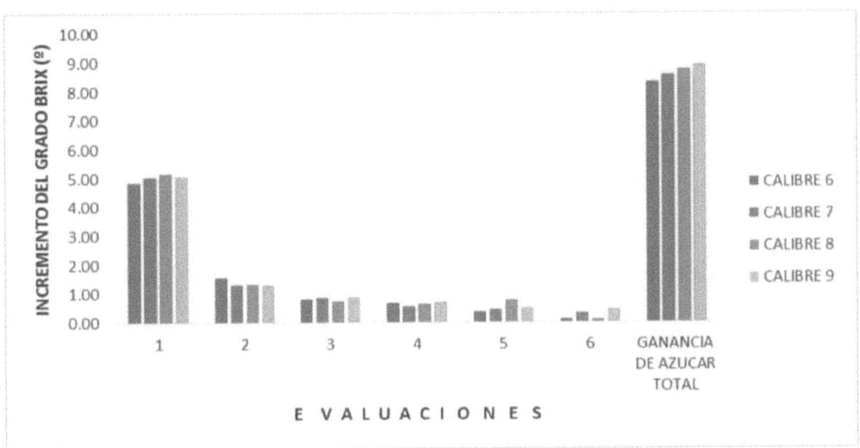

Figura 4.6. Increase of brix degree in mango fruits of Kent variety of four sizes in seven evaluations performed

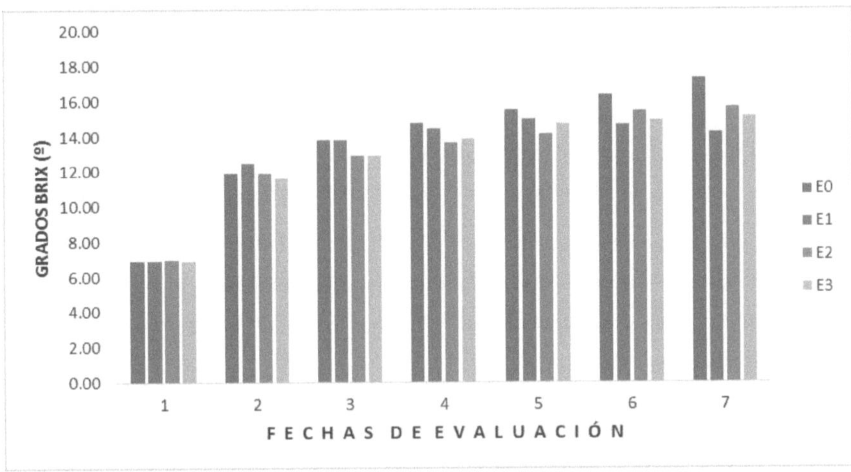

Figura 4.7. Brix degrees in mango fruit of Kent variety from four waxings in seven evaluations performed

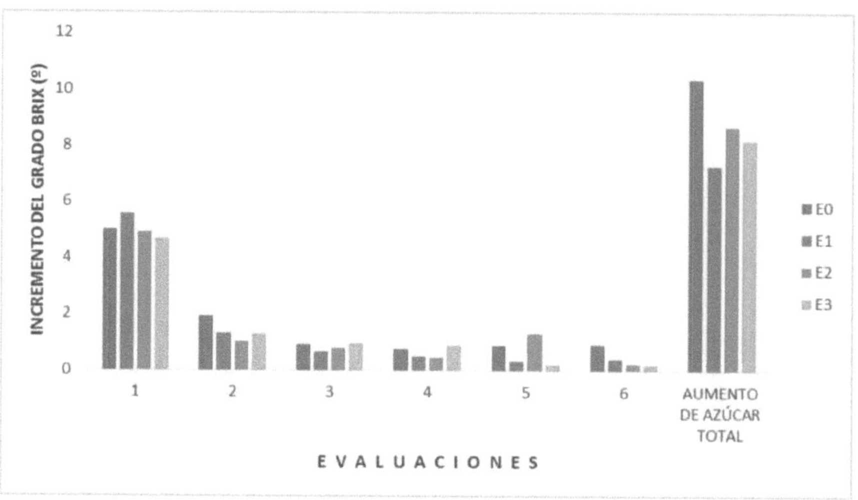

Figura 4.8. Increases of brix degree in mango fruits of Kent variety from four waxing in seven evaluations carried out.

4.3.- ACIDITY (%)

The percentage of acidity analyzed during the seven evaluations carried out on the Kent variety mango is shown in Tables 4.7a and 4.7b, and it can be seen that as far as the sizes are concerned, there were no statistically significant responses among them in all the evaluations carried out; the same occurred with the waxed varieties, which only had different statistical responses in the last two evaluations, one at the 0.05 level (sixth) and the other at the 0.01 level (seventh).

Neither was any significant statistical response detected in the interaction of the two factors studied, which indicates that they would act independently on this trait. The coefficients of variation ranged from 10.46% to 16.65%, values considered normal in this type of work. The records of the present observation can be found in annexes 16 to 22.

The summary of the Duncan0.05 tests performed on the percentage of acidity in the four mango sizes are shown in Table 4.8a, confirming conclusively the non-significance of the sizes on acidity, already shown in the respective ANVAS; there is a slight predominance in showing the lowest records of acidity, the size-9, which are the smallest mangoes, up to three times.

It is pertinent to point out that for this characteristic, the lowest acidity records would be the best or most adequate, unlike the brix degrees (°), where the highest values are the best. For a better understanding, see Figure 4.9.

Table 4.8b shows the variations in the percentage of acidity of the four sizes, showing that there were 11 cases of increased acidity and 13 cases of decreased acidity, with the first situation occurring in the initial evaluations, while the second was observed in the last evaluations. The greatest variation in acidity occurred in size-6, presenting a value of +0.331, which represented 262.7%, with respect to the original value; while the smallest variation +0.302, was presented by size-9, which are the fruits of lower weight, representing this value, 243.5%, with reference to the initial value. Figure 4.10 allows us to better visualize the details explained above.

To analyze the responses of the different types of waxes applied to the Kent variety mango, we will have to resort to Table 4.9a, where the results of the Duncan $0._{05}$ tests carried out are shown, from which we can deduce the following: Statistical significance was only detected between them in evaluations 6 and 7, but not in the first five, thus ratifying what has already been observed in the respective ANVAS (Table 4.7a and 4.7b).

As expected, the control treatment (without waxing), due to not having any protection, the acidity values presented were the highest in all the evaluations, with the exception of evaluation five. If we observe in more detail the acidity values in the three waxes, we will see that on four occasions, wax-3 (LUSTR 631) obtained the lowest values, and in the remaining three evaluations, Wax-1 (ECOWAX EXPORT MG) registered the lowest acidity values. Figure 4.11 shows the details given in the previous paragraph.

Finally, Table 4.9b shows the variations of acidity in the waxes in the seven evaluations carried out; it can be seen that the lowest increase in acidity, +0.293, was obtained by E3 (LUSTR 631), which, expressed as a percentage, corresponded to 238.2%, while the highest percentage (271.1%) was obtained by the control (E_0). See Figure 4.12.

Table 4.7.a. Summary of mean squares and statistical significance, in four evaluations of acidity (%) in mango fruits Var.

F. of Variation	G.L	EVALUATION			
		CM (1) 18 - 01 - 2016	CM (2) 25 - 01 - 2016	CM (3) 01 - 02 - 2016	CM (4) 08 - 02 - 2016
Blocks	3	0.00010 ns	0.00011 ns	0.0244 ns	0.0115 ns
Treatments	(15)				
Calibers	3	0.00006 ns	0.00028 ns	0.0223 ns	0.0102 ns
Waxing	3	0.00023 ns	0.00070 ns	0.0050 ns	0.0049 ns
C x E	9	0.00009 ns	0.00042 ns	0.0133 ns	0.0154 ns
Experimental error	45	0.00021	0.00044	0.0134	0.0093
Total	63	CV (%) = 11.78	16.65	15.37	12.48

ns = Not significant

* Statistical significance at 0.05 probability level.

** Statistical significance at 0.01 probability level.

Table 4.7.b. Summary of mean squares and statistical significance, in three evaluations of acidity (%) in mango fruit Var. Kent.

F. of Variation	G.L	EVALUATION		
		CM (5) 15 - 02 - 2016	CM (6) 22 - 02 - 2016	CM (7) 29 - 02 - 2016
Blocks	3	0.0039 ns	0.0019 ns	0.0039 ns
Treatments	(15)			
Calibers	3	0.0013 ns	0.0033 ns	0.0036 ns
Waxing	3	0.0058 ns	0.0172 *	0.0103 **
C x E	9	0.0004 ns	0.0028 ns	0.0008 ns
Experimental error	45	0.0032	0.0046	0.0021
Total	63	12.52	14.94	10.46

ns = Not significant

* Statistical significance at 0.05 probability level.

** Statistical significance at the 0.01 probability level.

Table 4.8.a. Summary of Duncan0.05 tests for acidity (%) in mango fruit, Var Kent, during seven evaluations, in four sizes.

CALIBER	EVALUATION						
	1	2	3	4	5	6	7
Caliber - 6 (646 - 700)	0.126 a	0.124 a	0.754 ab	0.776 a	0.445 a	0.469 a	0.457 a
Caliber - 7 (546 - 645)	0.121 a	0.123 a	0.764 ab	0.806 a	0.449 a	0.443 a	0.426 a
Caliber - 8 (481 - 545)	0.122 a	0.132 a	0.792 b	0.745 a	0.465 a	0.464 a	0.444 a
Caliber - 9 (426 - 480)	0.124 a	0.126 a	0.702 a	0.765 a	0.449 a	0.441 a	0.426 a

Note: Averages that have the same row are statistically equal, otherwise they are different.

Table 4.8.b. Variations of acidity (%) in mango fruit, Var Kent, in seven evaluations carried out, by size.

CALIBER	EVALUATION						ACIDITY VARIATION	%
	1	2	3	4	5	6		
Caliber - 6	- 0.002	+ 0.630	+ 0.022	- 0.331	+ 0.024	- 0.012	+ 0.331	262.7

Caliber - 7	+ 0.002	+ 0.641	+ 0.042	- 0.357	- 0.006	- 0.017	+ 0.305	252.1
Caliber - 8	+ 0.010	+ 0.660	- 0.047	- 0.280	- 0.001	- 0.020	+ 0.322	263.9
Caliber - 9	+ 0.002	+ 0.576	+ 0.063	- 0.316	- 0.008	- 0.015	+ 0.302	243.5
AVERAGE	+ 0.003	+ 0.627	+ 0.020	- 0.321	+ 0.002	- 0.016	+ 0.315	

Table 4.9.a. Summary of Duncan$_{0.05}$ tests for acidity (%) in mango fruit, Var Kent, during seven evaluations, by waxing .

WAXING	**EVALUATION**						
	1	2	3	4	5	6	7
E0	0.128 a	0.133 a	0.779 a	0.793 a	0.464 a	0.488 b	0.475 b
E1	0.119 a	0.123 a	0.742 a	0.774 a	0.435 a	0.474 b	0.431 a
E2	0.122 a	0.131 a	0.748 a	0.774 a	0.436 a	0.440 ab	0.432 a
E3	0.123 a	0.119 a	0.743 a	0.750 a	0.472 a	0.415 a	0.416 a

Note: Averages that have the same row are statistically equal, otherwise they are different.

Table 4.9.b. Variations of acidity (%) in mango fruit, Var Kent, in seven evaluations performed by waxing.

WAXING	**EVALUATION**						Increase Acidity	%
	1	2	3	4	5	6		
E0	+ 0.005	+ 0.646	+ 0.014	- 0.329	+ 0.024	- 0.012	+ 0.347	271.1
EI	+ 0.004	+ 0.619	+ 0.032	- 0.339	+ 0.039	- 0.043	+ 0.312	262.2
E2	+ 0.009	+ 0.617	+ 0.026	- 0.338	+ 0.004	- 0.008	+ 0.310	254.1
E3	- 0.004	+ 0.624	+ 0.007	- 0.278	- 0.057	+ 0.001	+ 0.293	238.2
AVERAGE	+ 0.004	+ 0.627	+ 0.020	- 0.321	+ 0.003	- 0.016	+ 0.316	256.4

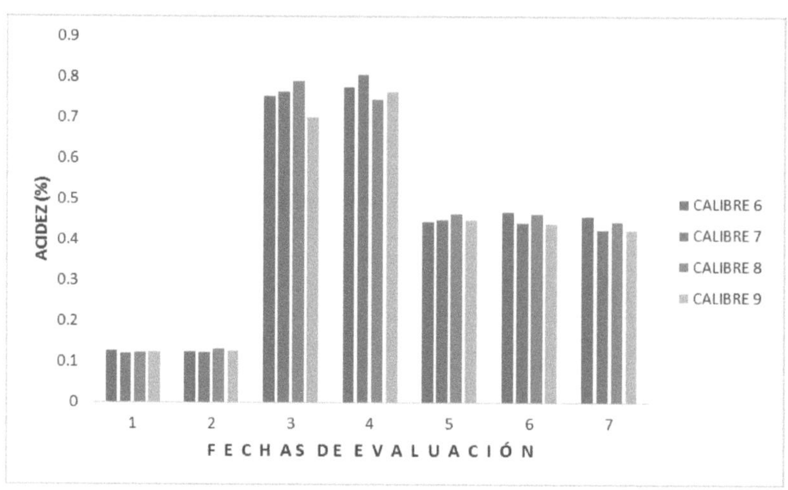

Figure 4.9. Fruit acidity in Var Kent mango of four sizes in seven evaluations (%).

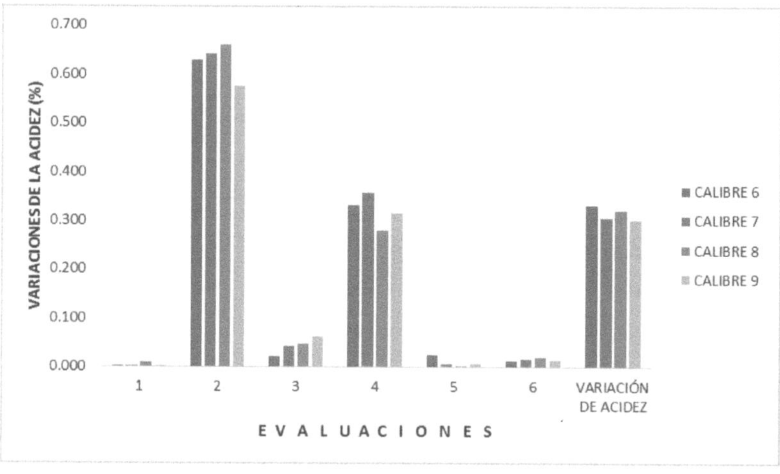

Figure 4.10. Variations of fruit acidity in mango Var. Kent of four sizes in seven evaluations (%).

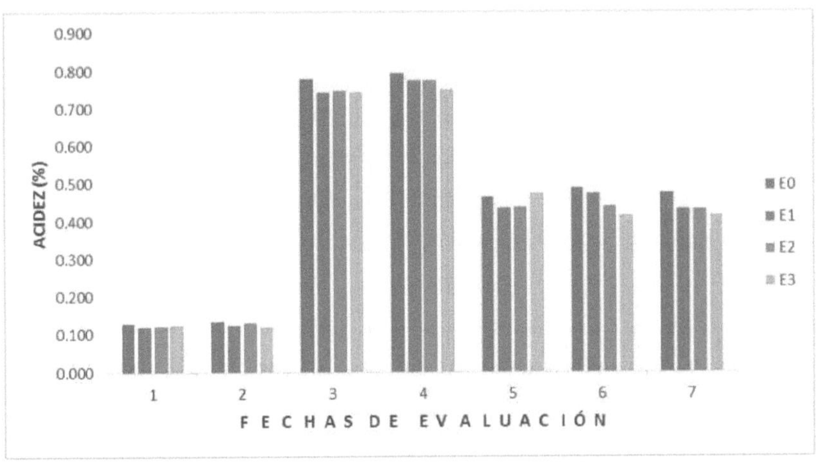

Figure 4.11. Acidity in mango fruit Var. Kent of four waxes in seven evaluations (%).

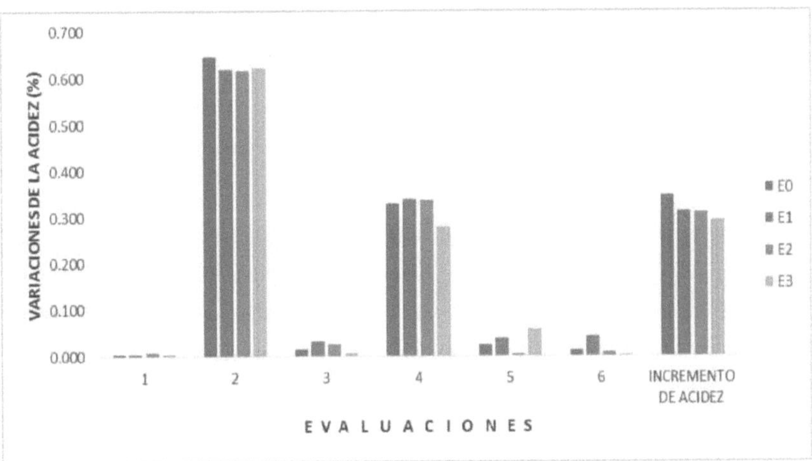

Figura 4.12. Variations of acidity in mango fruit Var. Kent from four waxings in seven evaluations (%).

4.4. VITAMIN "C" (mg. of Ascorbic Acid/100cc mango juice)

The values of the present observation are shown in annexes 23 to 29, and are expressed in mg of Ascorbic Acid/100 ml of juice. Tables 4.10a and 4.10b present the results of the mean squares and statistical significance of vitamin "C", evaluated on seven occasions, from which we can conclude the following:

No significant statistical differences were detected among the four mango sizes studied, with the exception of the fifth evaluation, where significance was found at the 0.05 level of probability. A similar response was found when the waxed mangoes were evaluated, but in this opportunity, no statistical differences were found in the seven evaluations performed.

Similarly, the C x E interaction did not register significant responses, which leads us to assume that these factors act independently on the vitamin "C" content of mango fruits.

The coefficients of variation ranged from 13.19% (fifth evaluation) to 24.21% (third evaluation), these values are considered quite acceptable and therefore give us confidence in the information presented.

The Duncan$0._{05}$ tests summarized for vitamin "C", as a consequence of the four calibers evaluated, are shown in table 4.11a, ratifying what was found in the respective ANVAS, and explained in the initial paragraphs. There is no pattern of response of the calibers, with respect to vitamin "C", observing only that in two opportunities each, calibers 7, 8 and 9, present the highest registers, and in only one case, it was obtained by caliber 6, within each evaluation. Therefore, it can be concluded from this table that waxing had no influence on the vitamin "C" parameter.

The highest value of vitamin "C" was 7.55 mg of ascorbic acid, obtained with the 9-gauge, in the penultimate evaluation, and the lowest values, 4.31 mg of ascorbic acid, were obtained with the 6-gauge, on two occasions (first and third evaluation). Figure 4.13 gives a more precise detail of what was explained above.

Table 4.11b records the variations of vitamin C content in the four sizes, showing that the smallest size fruit (size-9) obtained the greatest increase in vitamin C (+2.93 mg of ascorbic acid), with respect to the initial value, and the smallest increase was presented by size-6 (+2.26 mg of ascorbic acid), representing 67.8% and 52.4%, respectively, with respect to the initial value. See Figure 4.14.

On the other hand, Table 4.12a presents the summary of the Duncan$0._{05}$ tests, with respect to the waxes, showing that on three occasions each, the control (е0) and E3 (LUSTR 631) presented the highest values of this vitamin; and in only one case (fifth evaluation), the highest value was obtained by E2 (CITRASHINE ви3), with a record of 6.97 mg of ascorbic acid.

The highest value of vitamin "C" (7.37 mg) was obtained in the sixth evaluation, with в₃ (LUSTR 631), and the lowest value (4.01) was obtained in the first evaluation, with CITRASHINE EU3 (е2). Figure 4.15 illustrates better what has been explained above.

Finally, Table 4.12b shows the variations in vitamin "C" content with respect to the waxed products, showing that the greatest increases were obtained by waxed product 2 (CITRASHINE EU3), with a value of +3.18 mg of ascorbic acid, and the smallest increase, +2.21, was obtained by B-1 (ECOWAX EXPORT MG), these values representing 79.3% and 48.4%, respectively, in relation to the initial value. See figure 4.16.

Table 4.10.a. Summary of mean squares and statistical significance of vitamin C in mango fruit, Var. Kent, in four evaluations.

F. of Variation	G.L	EVALUATION			
		CM (1) 18 - 01 - 2016	CM (2) 25 - 01 - 2016	CM (3) 01 - 02 - 2016	CM (4) 08 - 02 - 2016
Blocks	3	0.3462 ns	0.8681 ns	1.9202 ns	1.7089 ns
Treatments	(15)				
Calibers	3	0.0137 ns	0.8479 ns	2.3850 ns	1.1013 ns
Waxing	3	1.4779 ns	0.2346 ns	0.6500 ns	1.1042 ns
C x E	9	0.3336 ns	0.5563 ns	3.0773 ns	1.2797 ns
Experimental error	45	0.7375	0.8804	3.0311	1.7971
Total	63	CV (%) = 19.83	17.38	24.21	20.62

Note: ns = Not significant

* Statistical significance at 0.05 probability level.

** Statistical significance at 0.01 probability level.

Table 4.10.b. Summary of mean squares and statistical significance of vitamin C in mango fruit Var. Kent, in three evaluations.

F. of Variation	G.L	EVALUATION		
		CM (5) 15 - 02 - 2016	CM (6) 22 - 02 - 2016	CM (7) 29 - 02 - 2016
Blocks	3	0.8482 ns	0.6244 ns	3.0038 ns
Treatments	(15)			
Calibers	3	2.8179 *	0.8584 ns	1.6629 ns
Waxing	3	0.7509 ns	0.6640 ns	0.4589 ns
C x E	9	1.1278 ns	0.7190 ns	1.5540 ns
Experimental error	45	0.7715	0.9420	1.0965
Total	63	13.19	13.48	14.98

Note: ns = Not significant

* Statistical significance at 0.05 probability level.

** Statistical significance at 0.01 probability level.

Table 4.11.a.Summary of Duncan's 0.05 tests for vitamin C in mango fruit Var Kent, in four sizes, during seven evaluations.

CALIBER	EVALUATION

	1	2	3	4	5	6	7
Caliber - 6 (646 - 700)	4.31 a	5.07 a	4.31 a	6.75 a	6.69 a	7.13 a	6.57 a
Caliber - 7 (546 - 645)	4.32 a	5.57 a	4.32 a	6.50 a	7.07 a	7.00 a	7.24 a
Caliber - 8 (481 - 545)	4.38 a	5.39 a	4.38 a	6.61 a	6.80 a	7.13 a	6.90 a
Caliber - 9 (426 - 480)	4.32 a	5.56 a	4.32 a	6.13 a	6.07 b	7.53 a	7.25 a

Note: Averages that have the same letter are statistically equal, otherwise they are different.

Table 4.11.b. Variations of vitamin C in mango fruit, Var Kent, in seven evaluations by size.

CALIBER	EVALUATION						VARIATION OF VITAMIN C	%
	1	2	3	4	5	6		
Caliber - 6	+ 0.76	- 0.76	+ 2.44	- 0.06	+ 0.44	- 0.56	+ 2.26	52.4
Caliber - 7	+ 1.25	- 1.25	+ 2.18	+ 0.57	- 0.07	+ 0.24	+ 2.92	67.6
Caliber - 8	+ 1.01	- 1.01	+ 2.23	+ 0.19	+ 0.33	- 0.23	+ 2.52	57.5
Caliber - 9	+ 1.24	- 1.24	+ 1.81	- 0.06	+ 1.46	- 0.28	+ 2.93	67.8
AVERAGE	+ 1.07	- 1.07	+ 2.17	+ 0.16	+ 0.54	- 0.21	+ 2.66	61.3

Table 4.12.a. Summary of Duncan 0.05 tests for vitamin C in mango Var Kent fruit during seven evaluations, by waxing.

WAXING	EVALUATION						
	1	2	3	4	5	6	7
E0	4.13 a	5.57 a	6.90 a	6.86 a	6.52 a	7.25 a	7.03 a
E1	4.57 a	5.37 a	7.30 a	6.24 a	6.63 a	6.90 a	6.78 a
E2	4.01 a	5.34 a	7.22 a	6.49 a	6.97 a	7.27 a	7.19 a
E3	4.61 a	5.30 a	7.35 a	6.40 a	6.51 a	7.37 a	6.96 a

Note: Averages that have the same letter are statistically equal, otherwise they are different.

Table 4.12.b. Variations of vitamin C in mango fruit Var Kent, in seven evaluations carried out by waxing.

WAXING	EVALUATION						Increase Acidity	%
	1	2	3	4	5	6		
E0	+ 1.44	+ 1.33	- 0.04	- 0.34	+ 0.73	- 0.22	+ 2.90	70.2
E1	+ 0.80	+ 1.93	+ 1.06	+ 0.39	+ 0.27	- 0.12	+ 2.21	48.4
E2	+ 1.33	+ 1.88	- 0.73	+ 0.48	+ 0.30	- 0.08	+ 3.18	79.3
E3	- 0.69	+ 2.05	- 0.95	+ 0.11	- 0.86	+ 0.41	+ 2.35	51.0

AVERAGE	+ 1.07	+ 1.80	- 0.70	+ 0.16	+ 0.54	- 0.21	+ 2.66	62.2

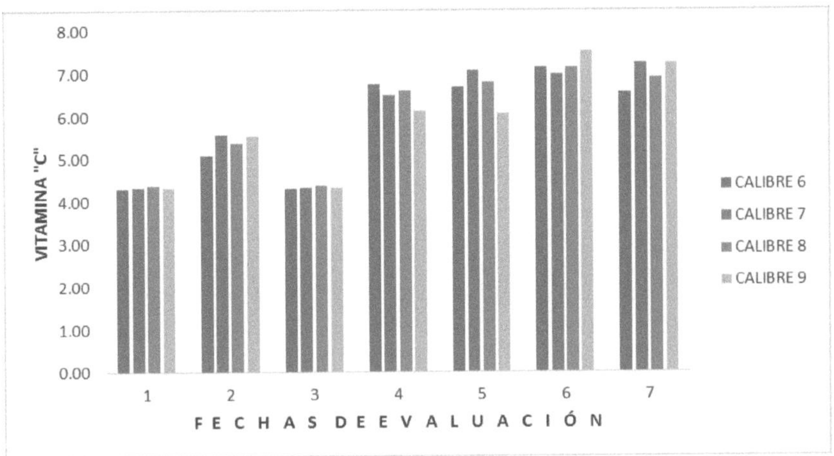

Figura 4.13.   Vitamin "C" in mango fruit Var. Kent of four different sizes in seven evaluations

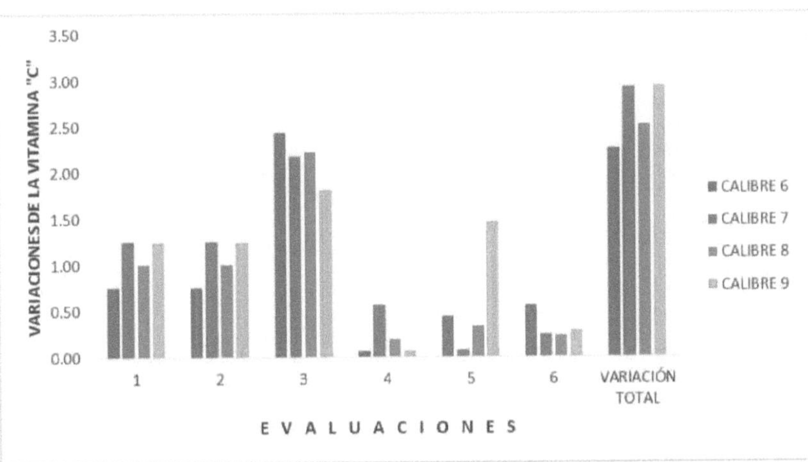

Figura 4.14.   Variations of vitamin "C" in mango fruits Var. Kent of four sizes in seven evaluations carried out.

Figura 4.15. Vitamin "C" in mango fruit Var. Kent from four waxing in seven evaluations performed

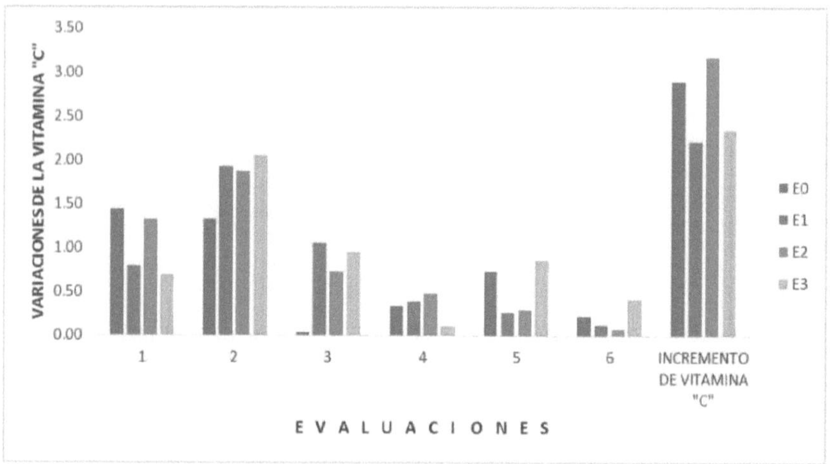

Figura 4.16. Variations of vitamin "C" in mango fruit Var. Kent of four waxing in seven evaluations carried out.

4.5. STRENGTH

In annexes 30 to 35, the original data of the present characteristic are presented, while in tables 4.13a and 4.13 b, the mean squares and statistical significance are summarized, showing the following: There are practically no statistically significant responses for the caliper factor, with the exception of that shown in evaluation 2 (P<0.05), while in the remaining evaluations, there was no significance. The opposite happened with the waxes, registering statistical significance in the seven evaluations, in six of them at the 0.01 level, and in the first, at the 0.05 level of probability; this confirms that the application of carnauba waxes and sucrose ester maintains the quality, contributing to preserve the firmness and reducing the weight losses of mango fruits in post-harvest. CACERES et.al (n.d).

The C x E interaction reported statistical differences in only one case (last evaluation), and it was at the 0.05 of probability, being this response an indicator that the two factors studied would work together on firmness in only this evaluation; appreciating that there is a tendency within each caliber to increase firmness when going from E0 (without waxing) to waxing E_1, E_2, and E_3, being practically the same response, but varying the magnitude of the changes, hence the significance detected (see table 4.15a).

The coefficients of variation (CV) ranged from 5.33% (third evaluation) to 10.74% (evaluation 6), which are considered to be low, thus giving us reliability to the information reported.

The summary of the $Duncan0_{.05}$ tests (Table 4.14a) confirms the findings of the respective ANVAS, showing that the smallest mangoes (sizes 8 and 9) obtained the highest firmness values, which are considered to be the best, obtaining in three cases each the highest values, and only in evaluation 4, size 6, presented the highest firmness value of 9.03 (see Figure 4.17).

Table 4.14b shows the decrease in firmness of the four sizes in all evaluations, with the 8-gauge showing the greatest total loss -10.02, which in percentage represented 58.2%.

Finally, the Duncan $0_{.05}$ tests (Table 4.15a) show that Waxing-2 (CITRASHINE EU 3) had the highest firmness values on five occasions; Waxing-3 (LUSTR 631) had a value of 8.06 in the last evaluation on only one occasion, and the control (E_o) had the highest firmness value of 17.52 in the initial evaluation.

Table 4.15b shows the decrease in firmness over the seven evaluations for each type of waxing, with no defined response pattern for this characteristic. All of the above can be corroborated by looking at Figures 4.18 to 4.2.

From the above it can be concluded that the treatments with wax have registered the highest firmness values; the best treatment being E3 (DECCO LUSTR 631), which at 42 days had a firmness of 8.06 KgF.

Table 4.13.a. Summary of mean squares and statistical significance of fruit firmness in mango Var. Kent in four evaluations.

F. of Variation	G.L	E V A L U A T I O N

		CM (1) 18 - 01 - 2016	CM (2) 25 - 01 - 2016	CM (3) 01 - 02 - 2016	CM (4) 08 - 02 - 2016
Blocks	3	0.9931 ns	1.4606 ns	0.2135 ns	0.6154 ns
Treatments	(15)				
Calibers	3	1.0177 ns	2.1460 *	0.1935 ns	0.1741 ns
Waxing	3	4.6570 ns	9.1860 **	1.9351 **	3.9833 **
C x E	9	0.5076 ns	1.2042 ns	0.0896 ns	0.1577 ns
Experimental error	45	1.1871	0.6224	0.4356	0.3440
Total	63	CV (%) = 6.42	5.88	5.33	6.57

Note: ns = Not significant

* Statistical significance at 0.05 probability level.

** Statistical significance at the 0.01 probability level.

Table 4.13.b. Summary of mean squares and statistical significance of firmness in mango fruit Var. Kent in three evaluations.

F. of Variation	G.L	**E V A L U A T I O N**		
		CM (5) 15 - 02 - 2016	CM (6) 22 - 02 - 2016	CM (7) 29 - 02 - 2016
Blocks	3	0.0501 ns	0.1628 ns	0.4743 ns
Treatments	(15)			
Calibers	3	0.0243 ns	0.2488 ns	0.1355 ns
Waxing	3	13.6047 **	12.2597 **	13.8934 **
C x E	9	0.1136 ns	0.2031 ns	0.4573 *
Experimental error	45	0.2043	0.5978	0.2158
Total	63	5.37	10.74	6.43

Note: ns = Not significant

* Statistical significance at 0.05 probability level.

** Statistical significance at the 0.01 probability level.

Table 4.14.a. Summary of Duncan's 0.05 tests for firmness in Var Kent mango fruit of four sizes in seven evaluations.

CALIBER	**E V A L U A T I O N**						
	1	2	3	4	5	6	7
Caliber - 6 (646 - 700)	16.98 a	12.96 b	12.31 a	9.03 a	8.39 a	7.25 a	7.14 a
Caliber - 7 (546	16.63 a	13.29 ab	12.45 a	8.79 a	8.38 a	7.03 a	7.26 a

| Caliber - 8 (481 - 545) | 17.21 a | 13.68 a | 12.49 a | 8.89 a | 8.46 a | 7.22 a | 7.19 a |
| Caliber - 9 (426 - 480) | 17.10 a | 13.74 a | 12.26 a | 8.98 a | 8.44 a | 7.32 a | 7.35 a |

Note: Averages that have the same letter are statistically equal, otherwise they are different.

Table 4.14.b. Decrease in firmness of Var Kent mango fruit by size in seven evaluations.

CALIBER	EVALUATION						DECREASE IN FIRMNESS	%
	1	2	3	4	5	6		
Caliber - 6	- 4.02	- 0.65	- 3.28	- 0.64	- 1.14	- 0.11	- 9.84	58.0
Caliber - 7	- 3.34	- 0.84	- 3.66	- 0.41	- 1.35	+ 0.23	- 9.37	56.3
Caliber - 8	- 3.53	- 1.19	- 3.60	- 0.43	- 1.24	- 0.03	- 10.02	58.2
Caliber - 9	- 3.36	- 1.48	- 3.28	- 0.54	- 1.12	+ 0.03	- 9.75	57.0
AVERAGE	- 3.56	- 1.04	- 3.46	- 0.51	- 1.21	+ 0.03	- 9.75	57.4

Table 4.15.a. Summary of Duncan$_{0.05}$ tests for firmness (Kgf) in Var Kent mango fruit, by waxing during seven evaluations.

WAXING	EVALUATION						
	1	2	3	4	5	6	7
E0	17.52 a	12.54 c	11.89 b	8.31 c	7.04 b	6.14 c	5.94 c
E1	16.75 a	13.66 b	12.38 a	8.76 b	8.95 a	6.84 b	7.21 b
E2	17.30 a	14.33 a	12.68 a	9.46 a	8.97 a	8.01 a	7.73 ab
E3	16.33 b	13.14 b	12.56 a	9.16 ab	8.69 a	7.83 a	8.06 a

Note: Averages that have the same letter are statistically equal, otherwise they are different.

Table 4.15.b. Decrease in firmness (Kgf) in Var Kent mango fruit by waxing, in seven evaluations carried out.

WAXING	EVALUATION						Total decrease	%
	1	2	3	4	5	6		
E0	- 4.98	- 0.65	- 3.58	- 1.27	- 0.90	- 0.20	- 11.58	66.1
E1	- 3.09	- 1.28	- 3.62	+ 0.19	- 2.11	+ 0.37	- 9.54	57.0
E2	- 2.97	- 1.65	- 3.22	- 0.49	- 0.96	- 0.28	- 9.57	55.3
E3	- 3.19	- 0.58	- 3.40	- 0.47	- 0.86	+ 0.23	- 8.27	50.6
AVERAGE	- 3.56	-1.04	- 3.46	- 0.44	- 1.21	+ 0.03		

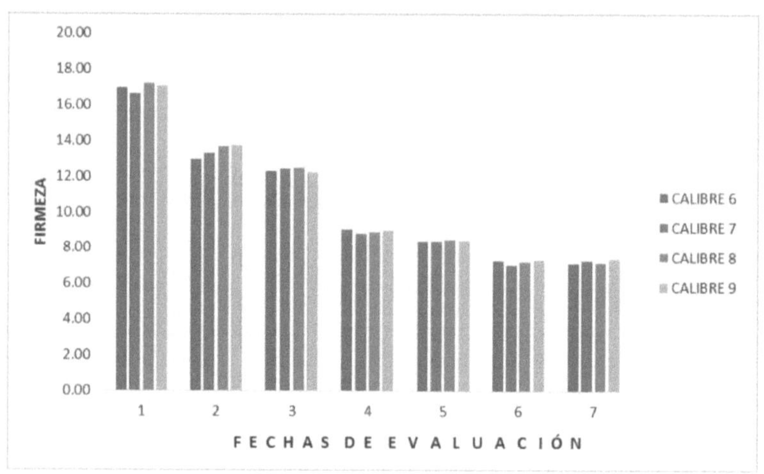

Figure 4.17. Firmness in mango fruit Var. Kent of four sizes in seven evaluations.

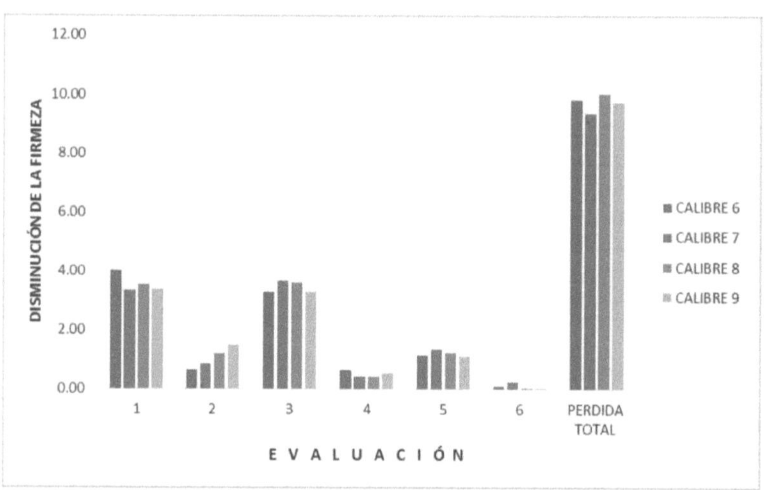

Figure 4.18. Decrease in firmness in mango fruit Var. Kent of four sizes in seven evaluations carried out.

Figura 4.19.   Firmness in mango fruit Var. Kent of four waxing in seven evaluations carried out.

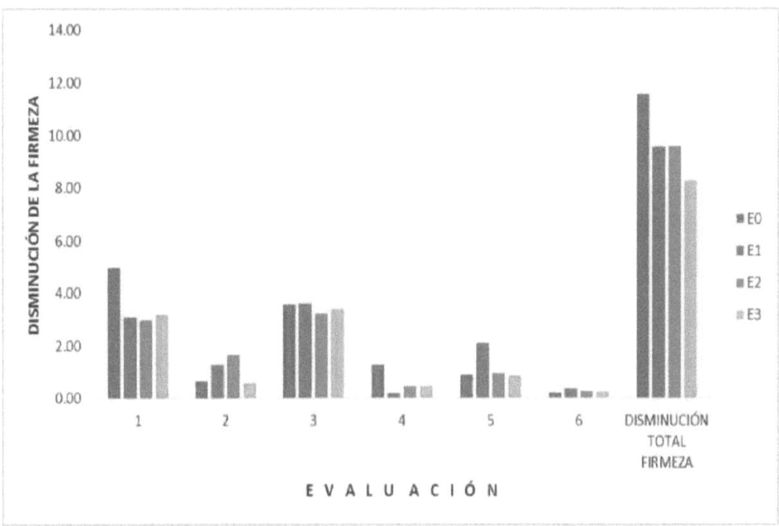

Figura 4.20.   Decrease of firmness in mango fruit Var. Kent of four waxing in seven evaluations carried out.

CHAPTER V

CONCLUSIONS

From the results obtained, considering the quality parameters evaluated in fruits of mango var. kent treated with hydrothermal treatment for 90 minutes, the following conclusions can be drawn:

1. The application of the different types of wax contributed to maintaining fruit quality because weight loss was lower when DECCO LUSTR 631 wax was applied (decreased 4.34% at 42 days after harvest), followed by the treatment with CITRASHINE EU 3 (4.45%), compared to 7.95% for the control treatment (WITHOUT WAX). Likewise, fruit firmness was higher, registering 8.06kgf for treatment E3 corresponding to DECCO LUSTR 631 wax, compared to 5.94kgf for the treatment without wax.

2. Other quality parameters evaluated were total soluble solids measured in brix degrees, as well as the percentage of acidity and vitamin C content; where the results show that the caliber and waxing factors had no significant effect in most of the evaluations, or their influence was minimal, being more difficult the action of waxes on the chemical constitution of the fruit of the Kent variety mango.

3. Regarding the size factor, it was found that the total weight loss is directly related to the size and weight of the fruit, i.e. larger and heavier fruits (size 6) lose more weight (34.4 g) and vice versa (size 9) lost a total of 26 g. However, taken as a percentage, there is no significant difference (5.12% and 5.69% respectively).

4. The Caliber by Waxing interaction was not statistically significant in any of the characteristics evaluated during the evaluations, so it is concluded that the factors act independently.

CHAPTER VI

RECOMMENDATIONS

Based on the conclusions obtained, the following is recommended:

1. The use of ECOWAX EXPORT MG, CITRASHINE EU 3 and DECCO LUSTR 631 waxes as post-harvest fruit preservation alternatives; considering that they reduce weight loss and maintain a greater firmness compared to the control treatment (WITHOUT WAX), in mango fruits var.

Kent until 42 days after harvest.

2. Evaluate the application of other types of waxy coatings, which could also include the evaluation of the performance of quality parameters such as internal coloration, external coloration, brightness and appearance of the fruit.

3. At present, different varieties of mango are exported from the Piura Region, such as EDWARD, HADEN, KEITT, ATAULFO, etc. Therefore, it is recommended to evaluate the behavior of each of these varieties against the application of waxes in shelf life.

CHAPTER VII

BIBLIOGRAPHIC REFERENCES

1. ACOSTA, G. 1988. The problems of post-harvest processes in fresh produce for export and their research needs. FUNDEAGRO. Chiclayo, Peru. p. 7.

2. AGROMAR INDUSTRIAL. 2015. Table of calibers. Piura. Peru.

3. ALARCON, G.1994. Methods of conservation, waxing, bagging and post-harvest refrigeration of mangoes. Piura Peru.

4. ATARAMA, 2008. Effect of the application of Raynox, calcium sulfate and hydrated lime, to control the insolation of mango Var. Kent in the department of Piura. Peru. Agronomist Thesis, Universidad Nacional de Piura. Piura, Peru.

5. BIOFRUIT SA. 2014. Quality control procedures .Piura-Peru.

6. CACERES, I; MULKAY, T; RODRiGUEZ, J; PAUMIER, A; SISINO, A; CASTRO-LÓPEZ, T; ALONSO, O; BANGO, G; AND GUTIÉRREZ, P. (n.d.). Influence of waxing and heat treatment on postharvest quality of mango. INSTITUTO DE INVESTIGACIONES EN FRUTICULTURA TROPICAL. ave.7ma. n°3005 e/ 30 y 32, Miramar. Playa. Havana City. Cuba. Email: icit@ceniai.inf.cu. Instituto Nacional de la Industria Azucarera Icinaz.

7. CALLE, D.1999. Comparative evaluation of shelf life in mango cultivation. TESIS UNP.150pp.

8. CORPORACION LITEC (n.d). Technical data sheet of citrashine eu 3 and decco lustr 631 waxes; Calle Los Aymaras 189 Monterrico, Lima 33-Perù. Available at www.litecperu.com

9. FRANCIOSI, R. 1992. Mango cultivation in Peru.ICE.Lima.Perù.69pp.

10. MENDOZA, J. 2005. Mangoes treatment in hot water 115°F (46.1°C); Piura - Peru.142 pàg.

11. MINISTRY OF AGRICULTURE, 2016. Available at the Office of Statistics and Informatics of the Regional Directorate of Agriculture; Piura - Peru. 120 pàg.

12. MINISTRY OF AGRICULTURE AND IRRIGATION 2017. Mango: Peruvian exports grow. technical profile No. 4. Available at: http://www.minagri.gob.pe/portal/analisis- economico/analisis-2017.

13. PUELLES, M. 2006. Effect of preservation methods of mango fruits (Mangifera indica L) cultivars Kent and haden, applied in post-harvest. Thesis Universidad Nacional de Piura. Piura-Peru.

14. SAN MARTIN 2012. Export Fruit Trees. Modular course. UNP. Piura-Peru. Pàg. 20

15. SAMSOM, J. A. 1991. Tropical Fruit Growing. First edition. Editorial UMUSA S.A., de CV Mexico. 509 pàg.

16. WOLFE, H.; E.OORT AND OTHERS, 1969 on mango cultivation in Peru. Ministry of Agriculture

and Fisheries. Dirección Regional de Investigaciones Agropecuarias. Boletin Tècnico nùmero 74. Lima - Perù.

17.   YAHÌA, E. M. 1992. Physiology and post-harvest technology of horticultural products. Editorial LIMUSA. Mexico. Pag.35.

ANNEXES

Annex 1. Initial fruit weight (g.) of Var Kent mango, in four sizes, for applying four types of waxing. Date: 18-012016 (evaluation 1).

\TRATAM. BLOCKS \	CALIBER - 6				CALIBER - 7				CALIBER - 8				CALIBER 9				TOTAL
	E_o	E_i	E_2	E_3	E_0	E_i	E_2	E_3	E_0	E_1	E_2	E_3	E_0	E_1	E_2	E_3	
I	669.3	668.0	684.0	673.3	626.0	586.0	571.3	596.7	521.3	522.0	532.7	526.0	441.3	462.0	452.0	455.3	8,987.2
II	664.7	677.3	672.7	655.3	604.7	590.0	614.7	604.7	518.7	522.7	528.7	521.3	464.7	469.7	450.0	456.0	9,015.9
III	675.3	670.7	672.7	675.3	606.7	586.7	608.7	612.7	529.3	536.0	519.3	528.7	449.3	451.7	451.3	456.0	9,030.4
IV	668.7	668.7	671.3	689.3	608.7	595.3	603.3	586.7	522.0	513.3	515.3	529.3	466.0	452.3	470.0	460.7	9,020.9
C X E	2678.0	2684.7	2700.7	2693.2	2446.1	2358.0	2398.0	2400.8	2091.3	2094.0	2096.0	2105.3	1821.3	1835.7	1823.3	1828.0	36,054.4
CALIBER	C6 = 10,756.6 672.3			л =	C7 = 9,602.9 600.2			-ï =	C8 = 8386.6 = 524.2			ϒ8	C9 = ï9 = 456.8			7308.3	ï0 = 563.4
WAXING	Eo = 9,036.7 = 564.8			7	Ei = 8,972.4 = 560.8			îi	E2 = 9018.0 = 563.6			ï2	E3 = 9027.3 = 564.2			73	

Annex 2. Fruit weight (g.) of Var Kent mango, in four sizes, after hydrothermal treatment. Date: 19 - 01 - 2016 (evaluation 2).

\ΓRATAM. BLOCKS X^x	CALIBER - 6				CALIBER - 7				CALIBER - 8				CALIBER 9				TOTAL
	E_o	E_i	E_2	E_3	E_0	E_1	E_2	E_3	E_0	E_1	E_2	E_3	E_0	E_1	E_2	E_3	
I	659.3	658.0	674.7	666.3	616.7	574.0	562.0	590.0	514.7	513.7	525.3	518.7	435.0	455.3	446.0	449.7	8,859.4
II	653.0	665.7	663.0	647.7	596.3	583.0	602.7	595.0	511.0	515.0	521.3	515.3	458.3	464.3	445.7	450.7	8,888.0
III	666.3	661.3	663.0	666.3	598.0	577.0	600.3	606.0	520.3	526.0	505.3	521.0	440.3	445.7	443.7	449.7	8,890.2
IV	660.3	658.7	660.0	680.0	600.7	585.7	593.0	577.3	514.3	508.3	508.3	521.3	456.3	446.8	463.3	454.0	8,888.3
C X E	2638.9	2643.7	2660.7	2660.3	2411.7	2319.7	2358.0	2368.3	2060.3	2063.0	2060.2	2076.3	1789.9	1812.1	1798.7	1804.1	35,525.9
CALIBER	C6 = 10,603.6 662.7			í6 =	C7 = 9457.7 591.1			Ï7=	C8 = 8259.8 516.2			í8 =	C9 = í9 = 450.3			7204.8	Д 55.1 =
WAXING	Eo = 8,900.8 ÆO = 556.3				E1 = 8,838.5 = 552.4			i1	E2 = 8,877.6 = 554.9			í2	E3 = 8,909.0 = 556.8			.U	

Annex 3. Fruit weight (g.) of Var Kent mango, in four sizes, with four types of waxing. Date: 25 - 01 - 2016(evaluation 3)

\TRATAM. BLOCKS\\	CALIBER - 6				CALIBER - 7				CALIBER - 8				CALIBER 9				TOTAL
	E_o	E_i	E_2	E_3	E_0	E_1	E_2	E_a	E_0	E_1	E_2	E_3	E_0	E_1	E_2	E_3	
I	648.7	650.7	669.3	661.0	604.3	568.7	556.0	586.0	505.3	507.3	521.0	514.0	428.7	449.3	441.7	445.7	8,757.7
II	641.0	659.7	657.7	642.0	584.7	577.0	598.0	589.3	500.7	510.0	517.3	510.7	450.3	459.7	442.3	446.3	8,786.7
III	654.3	655.0	657.3	660.3	586.3	569.3	596.3	601.7	509.0	519.7	510.3	517.0	430.0	440.7	440.0	446.0	8,793.2
IV	650.3	653.3	654.0	674.3	590.3	580.0	588.0	572.3	505.0	504.3	504.7	517.3	444.3	442.3	458.7	449.7	8,788.8
C X E	2594.3	2618.7	2638.3	2637.6	2365.6	2295.0	2338.3	2349.3	2020.0	2041.3	2053.3	2059.0	1753.3	1792.0	1782.7	1787.7	35,126.4

| CALIBER | C6 = 10,488.9 = 655.6 | 7- 584.3 | C7 = 9,348.2 | Ï7 = | C8 = 8,173.6 = 510.9 | 78 | C9 = 79 = 444.7 | 7,115.7 | Д = 548.9 |
| WAXING | Eo = 8,733.2 = 545.8 | X0 | E1 = 8,747.0 = 546.7 | 71 | E2 = 8,812.6 = 550.8 | 7; | E3 = 73 = 552.1 | 8,833.6 | |

Annex 4. Fruit weight (g.) of Var Kent mango, in four calibers, with four types of waxing. Date: 01 - 02 - 2016**(evaluation 4),**

\TRATAM. BLOCK\	CALIBER - 6				CALIBER - 7				CALIBER - 8				CALIBER 9				TOTAL
	Eo	E1	E2	E3	E0	E1	E2	E3	E0	E1	E2	E3	E0	E1	E2	E3	
I	642.0	647.7	665.3	658.0	597.0	563.7	553.3	583.0	500.3	504.7	517.7	511.0	424.7	446.0	439.7	444.7	8,698.8
II	633.3	655.3	654.0	639.0	578.3	575.0	594.3	586.0	495.3	506.7	513.3	508.0	445.0	458.0	440.0	445.0	8,726.5
III	646.3	651.3	654.3	657.3	578.3	566.0	593.0	598.7	502.3	516.3	508.0	513.7	423.3	437.0	437.0	441.7	8,724.5
IV	644.0	649.0	651.0	671.0	583.0	575.7	585.0	569.7	501.0	502.7	502.7	513.0	437.7	439.3	456.0	446.7	8,727.5
C X E	2,565.6	2,603.3	2624.6	2625.3	2336.6	2280.4	2325.6	2337.4	1998.9	2030.4	2041.7	2045.7	1730.7	1780.3	1772.7	1778.1	34,877.3
CALIBER	C6 = 10,418.8 Æ6 = 651.2				C7 = 9,280.0 Æ7 = 580.0				C8 = 8,116.7 Æ8 = 507.3				C9 = 7,061.8 Æ9 = 441.4				Д =544.95
WAXING	E0 = 8,631.8 Æ0 = 539.5				E1 = 8,694.4 Î1 = 543.4				E2 = 8,764.6 Î2 = 547.8				E3 = 8,786.5 549.2				Î3 =

Annex 5. Fruit weight (g.) of Var Kent mango, in four calibers, with four types of waxing. Date: 08 - 02 - 2016(evaluation 5).

\TRATAM. BLOCK\	CALIBER - 6				CALIBER - 7				CALIBER - 8				CALIBER 9				TOTAL
	Eo	Ei	E2	E3	E0	E1	E2	E3	E0	E1	E2	E3	E0	E1	E2	E3	
I	636.7	644.3	662.0	655.3	591.3	561.3	548.7	581.7	495.7	490.7	497.0	497.0	422.3	445.8	437.0	443.0	8,609.8
II	626.0	651.3	652.0	637.0	572.0	572.7	592.0	582.3	505.0	504.7	514.3	500.3	441.0	455.3	437.7	442.0	8,685.6
III	639.3	647.3	650.7	654.0	574.0	561.3	589.7	595.3	514.3	511.0	506.7	498.7	417.3	436.7	435.0	440.0	8,671.3
IV	640.0	646.3	648.3	668.0	578.0	572.7	582.0	566.0	507.7	504.7	511.0	511.0	430.7	437.9	453.7	444.7	8,702.7
C X E	2,542.0	2,589.2	2613.0	2614.3	2315.3	2268.0	2312.4	2325.3	2022.7	2011.1	2029.0	2007.0	1711.3	1775.7	1763.4	1769.7	34,669.4
CALIBER	C6 = 10,358.5 Æ6 = 647.4				C7 = 9,221.0 = 576.3				C8 = 8,069.8 Æ8 = 504.4				C9 = 7,020.1 Æ9 = 438.8				Д = 541.7
WAXING	E0 = 8,591.3 = 536.9				E1 = 8,644.0 = 540.3				E2 = 8,717.8 Î2 = 544.9				E3 = 8,716.3 544.8				Î3 =

Annex 6. Fruit weight (g.) of Var Kent mango, in four calibers, with four types of waxing. Date: 15 - 02 - 2016(evaluation 6)

\TRATAM. BLOCK\	CALIBER - 6				CALIBER - 7				CALIBER - 8				CALIBER 9				TOTAL
	Eo	Ei	E2	E3	E0	E1	E2	E3	E0	E1	E2	E3	E0	E1	E2	E3	
I	631.3	642.0	660.0	653.3	585.0	558.7	545.3	578.7	491.7	500.7	511.7	505.3	419.0	441.3	436.0	440.3	8,600.3
II	620.3	649.0	650.0	633.0	586.0	570.3	589.7	580.7	484.7	502.7	508.3	503.3	436.7	455.3	435.7	440.7	8,644.4
III	636.3	645.7	648.3	651.3	565.7	559.0	588.0	594.7	491.3	511.0	504.3	508.7	413.0	428.7	433.7	438.0	8,617.7
IV	635.3	643.7	645.3	665.0	572.0	570.7	579.7	564.3	492.0	499.0	498.3	508.7	427.3	431.0	451.7	443.0	8,627.0
C X E	2,523.2	2,580.4	2603.6	2602.6	2308.7	2258.7	2302.7	2318.4	1959.7	2013.4	2022.6	2026.0	1696.0	1754.3	1757.1	1762.0	34,489.4
CALIBER	C6 = 10,309.8 644.4				Ï6 = C7 = 9,188.5 = 574.3				Æ7 C8 = 8,021.7 Æ8 = 501.4				C9 = 6,969.4 Æ9 = 435.6				Λ = 538.90
WAXING	Eo = 8,487.6				E1 = 8,606.8				E2 = 8,686.0				E3 = 8,709.0				Î3 =

ÆO = 530.5				Î1 = 537.9				l2 = 542.9				544.3		

Annex 7. Fruit weight (g.) of Var Kent mango, in four calibers, with four types of waxing. Date: 22 - 02 - 2016(evaluation 17

\TRATAM. BLOCK\	CALIBER - 6				CALIBER - 7				CALIBER - 8				CALIBER 9				TOTAL
	Eo	E1	E2	E3	E0	E1	E2	E3	E0	E1	E2	E3	E0	E1	E2	E3	
I	6.25.7	639.3	656.0	649.3	578.3	556.3	541.7	576.7	486.0	497.7	508.0	501.7	414.3	440.0	433.3	438.3	8,542.6
II	613.3	645.3	647.0	629.0	560.3	567.7	585.3	576.3	579.0	500.7	506.0	500.7	431.7	451.7	434.0	438.7	8,666.7
III	628.0	642.3	644.7	648.0	558.7	555.3	584.3	591.7	483.7	507.7	502.3	505.3	407.0	424.7	431.3	435.0	8,550.0
IV	630.0	639.7	640.7	664.3	564.7	568.0	576.7	561.3	486.3	497.0	496.7	505.0	418.3	427.7	449.7	440.0	8,566.1
C X E	2,497.0	2,566.6	2588.4	2590.6	2262.0	2247.3	2288.0	2306.0	2035.0	2003.1	2013.0	2012.7	1671.3	1744.1	1748.3	1752.0	34,325.4
CALIBER	C6 = 10,242.6 Æ6 = 640.2				C7 = 9,103.3 Æ7 = 568.9				C8 = 8,063.8 Æ8 = 503.9				C9 = 6,915.7 Æ9 = 432.2				Д = 536.3
WAXING	E0 = 8,465.3 Îo = 529.1				E1 = 8,561.1 Î1 = 535.1				E2 = 8,637.7 l2 = 539.9				E3 = 8,661.3 l3 = 541.3				

Annex 8. Fruit weight (g.) of Var Kent mango, in four calibers, with four types of waxing. Date: 29 - 02 - 2016(evaluation 8)

\TRATAM. BLOCK\	CALIBER - 6				CALIBER - 7				CALIBER - 8				CALIBER 9				TOTAL
	Eo	Ei	E2	E3	E0	E1	E2	E3	E0	E1	E2	E3	E0	E1	E2	E3	
I	622.7	637.7	654.7	647.7	575.0	554.7	540.0	575.7	483.3	497.0	506.7	500.3	412.3	439.0	432.7	437.0	8,516.5
II	609.7	643.7	645.3	627.0	557.0	566.0	583.7	574.7	476.0	500.0	504.3	499.3	429.0	450.7	432.7	437.7	8,536.8
III	624.3	640.7	643.0	646.3	555.7	554.3	583.0	590.7	480.7	506.7	501.3	504.0	404.0	423.3	430.3	433.7	8,522.0
IV	627.3	638.7	639.7	659.3	561.7	567.0	575.0	559.7	484.3	495.0	495.7	503.7	416.0	426.3	448.7	439.0	8,537.1
C X E	2,484.0	2,560.8	2582.7	2580.3	2249.4	2242.0	2281.7	2300.8	1924.3	1998.7	2008.0	2007.3	1661.3	1739.3	1744.4	1747.4	34,112.4
CALIBER	C6 = 10,207.8 -Ï6 = 637.9				C7 = 9,073.9 Æ7 = 567.1				C8 = 7,938.3 Æ8 = 496.1				C9 = 6,892.4 -Ï9 = 430.8				Λ = 533.00
WAXING	Eo = 8,319.0 -Ï0 = 519.9				E1 = 8,540.8 Æ1 = 533.8				E2 = 8,616.8 -Ï2 = 538.5				E3 = 8,635.8 Æ3 = 539.7				

Annex 9. Brix degrees in Var Kent mango fruit, in four sizes, with four types of waxing. Date: 18 - 01 - 2016(evaluation 1)

\TRATAM. BLOCK\	CALIBER - 6				CALIBER - 7				CALIBER - 8				CALIBER 9				TOTAL
	Eo	E1	E2	E3	E0	E1	E2	E3	E0	E1	E2	E3	E0	E1	E2	E3	
I	6.9	7.0	7.3	6.8	7.2	6.7	7.1	6.9	6.8	7.1	6.9	6.6	7.0	6.6	6.7	6.9	110.5
II	7.1	6.8	6.8	6.9	7.0	6.8	7.0	6.9	6.9	7.0	7.0	6.9	6.7	6.7	7.1	7.0	110.6
III	7.0	6.9	6.9	7.2	6.5	7.0	6.8	6.9	6.8	7.1	6.8	7.0	7.0	7.1	6.9	6.8	110.7
IV	6.6	6.8	7.2	7.0	7.1	7.0	6.9	6.8	6.9	7.2	7.0	6.9	6.9	7.0	7.1	7.0	111.4
C X E	27.6	27.5	28.2	27.9	27.8	27.5	27.8	27.5	27.4	28.4	27.7	27.4	27.6	27.4	27.8	27.7	443.2
CALIBER	C6 = 111.2 Я6 = 6.95				C7 = 110.6 Я7 = 6.91				C8 = 110.9 -Ï8 = 6.93				C9 = 110.5 X9 = 6.91				Д = 6.93
WAXING	Eo = 110.4 Xo = 6.90				E1 = 110.8 7i = 6.93				E2 = 111.5 -T2 = 6.97				E3 = 110.5 l3 = 6.91				

Brix degrees in Var Kent mango fruit, in four sizes, with four types of waxing. Date: 25 - 01 - 2016 (evaluation 2)

\TRATAM. BLOCKS\\	CALIBER - 6				CALIBER - 7				CALIBER - 8				CALIBER 9				TOTAL
	Eo	E1	E2	E3	EG	E1	E2	E3	EG	E1	E2	E3	EG	E1	E2	E3	
I	10.7	13.0	11.5	10.1	12.0	11.8	12.0	12.0	13.0	12.4	12.0	11.5	12.2	13.1	12.2	11.0	190.5
II	12.0	12.0	12.0	11.0	12.1	12.5	11.7	11.5	11.8	12.0	12.2	11.9	11.8	11.6	11.6	11.5	189.2
III	11.9	12.9	11.8	11.2	11.5	12.1	11.6	11.8	11.5	12.7	11.9	12.0	12.0	12.8	12.0	12.0	191.7
IV	12.2	12.5	12.1	12.0	12.1	12.7	11.8	12.3	12.0	13.0	11.5	12.1	11.7	12.5	12.0	11.8	194.3
C X E	46.8	50.4	47.4	44.3	47.7	49.1	47.1	47.6	48.3	50.1	47.6	47.5	47.7	50.0	47.8	46.3	765.7
CALIBER	C6 = 188.9 11.80		л =		C7 = S7= 11.96			191.5	C8 = 193.5 12.09			E =	C9 = 191.8 11.98			Y9 =	Д = 11.96
WAXING	EG = 190.5 = 11.91		Æü	E1 = Ж1 = 12.48			199.6	E2 = 189.9 = 11.87			я2	E3 = 185.7 11.61			S3 =		

Brix degrees in Var Kent mango fruit, in four sizes, with four types of waxing. Date: 01 - 02 - 2016 (evaluation 3)

\TRATAM. BLOCK\	CALIBER - 6				CALIBER - 7				CALIBER - 8				CALIBER 9				TOTAL
	Eo	E1	E2	E3	EQ	E1	E2	E3	EQ	E1	E2	E3	EQ	E1	E2	E3	
I	14.1	14.0	12.7	13.0	13.8	14.0	13.0	13.0	14.3	13.7	13.1	13.3	13.5	13.5	12.7	13.0	214.7
II	14.0	13.9	13.0	12.6	13.5	13.6	12.8	12.5	13.9	13.9	12.8	12.9	14.0	14.0	13.0	12.5	212.9
III	13.8	13.8	13.1	12.9	13.8	13.8	13.0	12.8	13.6	14.0	13.0	13.1	13.5	14.1	12.6	12.8	213.7
IV	13.5	13.5	12.9	13.0	14.0	13.5	12.5	12.9	14.0	13.5	12.9	13.0	13.8	14.0	12.8	13.0	212.8
C X E	55.4	55.2	51.7	51.5	55.1	54.9	51.3	51.2	55.8	55.1	51.8	52.3	54.8	55.6	51.1	51.3	854.1
CALIBER	C6 = 213.8 13.36		76 =		C7 = 77= 13.28			212.5	C8 = 215.0 13.44			78 =	C = 212.8 13.30			7G =	Д = 13.35
WAXING	EQ = 221.1 = 13.82		Æü	E1 = 220.8 = 13.80			71	E2 = 205.9 = 12.87			72	E3 = 206.3 12.89			73 =		

Brix degrees in Var Kent mango fruit, in four sizes, with four types of waxing. Date: 08 - 02 - 2016 (evaluation 4)

\TRATAM. BLOCKSx	CALIBER - 6				CALIBER - 7				CALIBER - 8				CALIBER 9				TOTAL
	Eo	E1	E	E3	Eo	E1	E2	E3	Eo	E1	E2	E3	Eo	E1	EG	E3	
I	14.8	14.4	13.9	13.7	14.7	14.5	13.8	13.9	14.6	14.4	13.6	13.8	15.1	14.6	13.5	14.0	227.3
II	14.7	14.4	13.6	13.6	14.5	14.4	13.7	13.8	14.8	14.5	13.7	14.0	14.8	14.5	13.6	13.8	226.4
III	14.6	14.3	13.5	13.8	14.6	14.3	13.8	14.0	15.0	14.6	13.5	13.9	14.7	14.4	13.7	13.7	226.4
IV	14.9	14.4	13.8	14.0	14.7	14.4	13.5	13.8	14.7	14.5	13.6	13.7	14.8	14.6	13.8	13.9	227.1
C X E	59.0	57.5	54.8	55.1	58.5	57.6	54.8	55.5	59.1	58.0	54.4	55.4	59.4	58.1	54.6	55.4	907.2
CALIBER	C6 = 226.4 14.15		76 =		C7 = 226.4 = 14.15			77	C8 = 226.9 14.18			78 =	C9 = 227.5 14.21			7. =	Д =14.175
WAXING	EG = 236.0 = 14.75		7o	E1 = 231.2 = 14.45			71	EG = 218.6 = 13.66			7;	E3 = 221.4 13.83			73 =		

Brix degrees in Var Kent mango fruit, in four sizes, with four types of waxing. Date: 15 - 02 - 2016 (evaluation 5)

\TRATAM. BLOCKSx	CALIBER - 6				CALIBER - 7				CALIBER - 8				CALIBER 9				TOTAL
	Eo	E1	E2	E3	E0	E1	E2	E3	E0	E1	E2	E3	E0	E1	E2	E3	

	Eo	E1	E2	E3	E0	E1	E2	E3	E0	E1	E2	E3	E0	E1	E2	E3	TOTAL
I	14.9	15.5	14.5	14.0	15.8	14.8	13.7	15.0	15.9	14.8	14.0	14.2	15.6	15.0	14.4	15.1	237.2
II	15.8	15.0	14.4	15.1	14.8	14.7	14.0	14.8	15.5	15.0	13.8	15.0	15.9	15.4	14.2	14.9	238.3
III	15.5	14.9	14.2	15.0	14.8	15.4	14.2	15.1	15.6	15.2	14.5	14.0	15.6	14.9	14.0	15.0	237.9
IV	15.6	14.3	14.0	14.7	15.0	15.0	14.3	14.0	15.7	14.8	14.0	15.0	15.8	14.5	13.9	14.5	235.1
C X E	61.8	59.7	57.1	58.8	60.4	59.9	56.2	58.9	62.7	59.8	56.3	58.2	62.9	59.8	56.5	59.5	948.5
CALIBER	C6 = 237.4 14.83			7β =	C7 = 235.4 14.71			77=	C8 = 237.0 14.81			78 =	C9 = 238.7 14.92			7P =	Д = 14.82
WAXING	Eo = 247.8 = 15.49			-ïo	E1 = 239.2 = 14.95			71	E2 = 226.1 = 14.13			72	E3 = 235.4 14.71			73 =	

Brix degrees in Var Kent mango fruit, in four sizes, with four types of waxing. Date: 22 - 02 - 2016 (evaluation 6)

\TRATAM.	CALIBER - 6				CALIBER - 7				CALIBER - 8				CALIBER 9				TOTAL
BLOCKSx	Eo	E1	E2	E3	E0	E1	E2	E3	E0	E1	E2	E3	E0	E1	E2	E3	
I	15.2	13.9	15.5	15.0	15.1	14.7	14.6	13.8	15.0	14.9	16.2	15.5	18.0	14.1	14.8	15.3	241.6
II	16.0	14.0	14.8	14.0	16.5	13.8	15.0	15.0	16.5	14.5	15.8	14.8	15.5	15.0	16.0	14.0	241.2
III	16.5	14.5	16.0	15.5	17.0	15.0	16.0	14.5	17.8	15.0	15.5	15.0	16.0	14.8	15.0	16.0	250.1
IV	15.5	15.0	15.5	16.0	16.8	14.9	15.8	14.0	17.5	14.8	14.9	15.8	16.8	15.0	15.5	14.5	248.3
C X E	63.2	57.4	61.8	60.5	65.4	58.4	61.4	57.3	66.8	59.2	62.4	61.1	66.3	58.9	61.3	59.8	981.2
CALIBER	C6 = 242.9 15.18			X6 =	C7 = 242.5 77= 15.16				C8 = 249.5 15.59			78 =	C9 = 246.3 15.39			79 =	Д = 15.33
WAXING	Eo = 261.7 16.36			7 =	E1 = 233.9 = 14.62			71	E2 = 246.9 = 15.43			72	E3 = 238.7 14.92			73 =	

Brix degrees of Var Kent mango fruit, in four sizes, with four types of waxing. Date: 29 - 02 - 2016 (evaluation 7)

\TRATAM.	CALIBER - 6				CALIBER - 7				CALIBER - 8				CALIBER 9				TOTAL
BLOCKSx	Eo	E1	E2	E3	E0	E1	E2	E3	E0	E1	E2	E3	E0	E1	E2	E3	
I	17.1	14.0	13.9	15.1	17.0	14.8	13.5	14.0	16.5	13.9	15.0	14.8	18.5	15.0	17.0	14.3	244.4
II	17.0	13.5	16.0	14.5	18.0	15.0	15.0	15.0	19.0	14.0	16.0	15.0	18.0	14.5	16.8	15.5	252.8
III	15.9	15.0	14.5	16.0	16.5	13.5	16.0	14.5	18.0	14.5	14.5	16.0	16.9	13.5	16.0	16.0	247.3
IV	16.0	14.0	17.0	14.8	18.0	14.5	16.5	16.0	17.0	13.8	17.0	15.8	17.0	13.9	15.9	14.5	251.7
C X E	66.0	56.5	61.4	60.4	69.5	57.8	61.0	59.5	70.5	56.2	62.5	61.6	70.4	56.9	65.7	60.3	996.2
CALIBER	C6 = 244.3 15.27			7e =	C7 = 247.8 77= 15.49				C8 = 250.8 15.68			78 =	C9 = 253.3 15.83			79 =	Д = 15.57
WAXING	E0 = 276.4 17.28			7 =	E1 = 227.4 = 14.21			71	E2 = 250.6 = 15.66			72	E3 = 241.8 15.11			7s =	

Acidity in mango fruit Var. Kent, in four sizes, with four types of waxing. Date: 18 - 01 - 2016(evaluation 1)

\TRATAM.	CALIBER - 6				CALIBER - 7				CALIBER - 8				CALIBER 9				TOTAL
BLOCKSx	Eo	E1	E2	E3	E0	E1	E2	E3	E0	E1	E2	E3	E0	E1	E2	E3	
I	0.145	0.140	0.131	0.100	0.120	0.102	0.110	0.125	0.120	0.110	0.110	0.130	0.120	0.144	0.120	0.110	1.937
II	0.130	0.120	0.110	0.140	0.130	0.140	0.120	0.100	0.120	0.130	0.120	0.140	0.130	0.120	0.124	0.105	1.979
III	0.130	0.110	0.145	0.120	0.130	0.110	0.135	0.120	0.150	0.115	0.130	0.120	0.140	0.110	0.130	0.130	2.025
IV	0.140	0.100	0.120	0.130	0.110	0.100	0.130	0.150	0.110	0.140	0.100	0.110	0.120	0.110	0.120	0.143	1.933

C X E	0.545	0.470	0.506	0.490	0.490	0.452	0.495	0.495	0.500	0.495	0.460	0.500	0.510	0.484	0.494	0.488	7.874
CALIBER	C6 = 2.011		X6 = 0.126		C7 = 1.932		X7= 0.121		C8 = 1.955		X8 = 0.122		C9 = 1.976		Æ9 = 0.124		Д = 0.123
WAXING	E0 = 2.045		Ĩ0 = 0.128		E1 = 1.901		Ï1 = 0.119		E2 = 1.955		2 = 0.122		E3 = 1.973		Ĩ3 = 0.123		

Acidity in Var Kent mango fruit, in four sizes, with four types of waxing. Date: 25-01 - 2016 (evaluation 2).

\TRATAM.	CALIBER - 6				CALIBER - 7				CALIBER - 8				CALIBER 9				TOTAL
BLOCKSx	E0	E1	E2	E3	E0	E1	E2	E3	E0	E1	E2	E3	E0	E1	E2	E3	
I	0.157	0.122	0.131	0.096	0.147	0.131	0.090	0.112	0.141	0.147	0.109	0.102	0.154	0.096	0.157	0.115	2.007
II	0.122	0.115	0.128	0.147	0.115	0.096	0.154	0.128	0.154	0.157	0.115	0.128	0.128	0.154	0.115	0.090	2.046
III	0.154	0.109	0.115	0.128	0.128	0.115	0.147	0.102	0.128	0.096	0.154	0.147	0.115	0.147	0.160	0.122	2.067
IV	0.128	0.096	0.122	0.115	0.147	0.134	0.102	0.122	0.115	0.122	0.147	0.157	0.096	0.128	0.147	0.093	1.971
C X E	0.561	0.442	0.496	0.486	0.537	0.476	0.493	0.464	0.538	0.522	0.525	0.534	0.493	0.525	0.579	0.420	8.091
CALIBER	C6 = 1.985		Æ6 = 0.124		C7 = 1.970		77= 0.123		C8 = 2.119		78= 0.132		C9 = 2.017		Æ9 = 0.126		Д =0.126
WAXING	E0 = 2.129		Æo = 0.133		E1 = 1.965		71 = 0.123		E2 = 2.093		72 = 0.131		E3 = 1.904		7э = 0.119		

Acidity of Var Kent mango fruit, in four sizes, with four types of waxing. Date: 01 - 02 - 2016 (evaluation 3)

\TRATAM.	CALIBER - 6				CALIBER - 7				CALIBER - 8				CALIBER 9				TOTAL
BLOCKSx	E0	E1	E2	E3	EQ	E1	E2	E3	EQ	E1	E2	E3	EQ	E1	E2	E3	
I	0.845	0.646	0.928	0.538	0.640	0.736	0.781	0.838	0.928	0.544	0.627	0.800	0.653	0.723	0.461	0.589	11.277
II	0.800	0.608	0.845	0.736	0.832	0.915	0.672	0.883	0.800	0.858	0.736	0.890	0.787	0.864	0.544	0.909	12.679
III	0.704	0.902	0.678	0.742	0.736	0.851	0.659	0.544	0.864	0.672	0.915	0.749	0.659	0.614	0.800	0.736	11.825
IV	0.768	0.608	0.832	0.877	0.896	0.851	0.794	0.602	0.915	0.832	0.800	0.736	0.640	0.646	0.896	0.717	12.410
C X E	3.117	2.764	3.283	2.893	3.104	3.353	2.906	2.867	3.507	2.906	3.078	3.175	2.739	2.847	2.701	2.951	48.191
CALIBER	C6 = 12.057		X6 = 0.754		C7 = 12.230 Æ7 = 0.764				C8 = 12.666 0.792		E =		C9 = 11.238 = 0.702		Æ9		V. = 0.753
WAXING	EQ = 12.467 = 0.779		Ĩ0		E1 = 11.870 = 0.742				E2 = 11.968 = 0.748 i1				E3 = 11.886 0.743 Î2				Î3 =

Acidity in Var Kent mango fruit, in four sizes, with four types of waxing. Date: 08 - 02 - 2016 (evaluation 4)

\TRATAM.	CALIBER - 6				CALIBER - 7				CALIBER - 8				CALIBER 9				TOTAL
BLOCKSx	E0	E1	E2	E3	EQ	E1	E2	E3	EQ	E1	E2	E3	EQ	E1	E2	E3	
I	0.806	0.608	0.864	0.627	0.902	0.890	0.787	0.896	0.678	0.634	0.717	0.595	0.672	0.627	0.774	0.762	11.839
II	0.742	0.749	0.685	0.902	0.685	0.742	0.620	0.890	0.928	0.877	0.800	0.659	0.774	0.915	0.736	0.864	12.568
III	0.806	0.685	0.806	0.832	0.806	0.883	0.634	0.800	0.787	0.902	0.749	0.634	0.800	0.659	0.832	0.608	12.223
IV	0.883	0.902	0.890	0.621	0.883	0.813	0.736	0.922	0.672	0.698	0.858	0.730	0.864	0.800	0.896	0.659	12.827
C X E	3.237	2.944	3.245	2.982	3.276	3.328	2.777	3.508	3.065	3.111	3.124	2.618	3.110	3.001	3.238	2.893	49.457
CALIBER	C6 = 12.408 0.776		X6 =		C7 = 12.889 Æ7 = 0.806				C8 = 11.918 0.745		E =		C9 = 12.242 0.765		Æ9 =		Д = 0.773
WAXING	EQ = 12.688 = 0.793		Ĩ0		E1 = 12.384 Æ1 = 0.774				E2 = 12.384 = 0.774 Î2				E3 = 12.001 0.750		Î3 =		

Acidity in Var Kent mango fruit, in four sizes, with four types of waxing. Date: 15 - 02 - 2016

(evaluation 5)

\TRATAM.	CALIBER - 6				CALIBER - 7				CALIBER - 8				CALIBER 9				TOTAL
BLOCKSx	Eo	E1	E2	E3	EQ	E1	E2	E3	EQ	E1	E2	E3	EQ	E1	E2	E3	
I	0.454	0.429	0.480	0.416	0.580	0.397	0.390	0.416	0.474	0.435	0.550	0.397	0.403	0.429	0.365	0.403	7.018
II	0.448	0.320	0.454	0.512	0.416	0.448	0.480	0.480	0.474	0.429	0.435	0.480	0.480	0.416	0.512	0.570	7.354
III	0.544	0.474	0.371	0.448	0.442	0.544	0.461	0.512	0.506	0.512	0.371	0.525	0.480	0.454	0.403	0.480	7.527
IV	0.448	0.435	0.403	0.480	0.403	0.397	0.371	0.454	0.442	0.403	0.442	0.570	0.435	0.442	0.493	0.416	7.034
C X E	1.894	1.658	1.708	1.856	1.841	1.786	1.702	1.862	1.896	1.779	1.798	1.972	1.798	1.741	1.773	1.869	28.933
CALIBER	C6 = 7.116 0.445		76	=C7 = 7.191 0.449			77=	C8 = 7.445 0.465			78	=C9 = 7.181 0.449			79	=fG 0.452	=
WAXING	E0 = 7.429 0.464		7o	=E1 = 6.964 = 0.435			71	E2 = 6.981 0.436			72	=E3 = 7.559 0.472			73		=

Annex 21. Acidity in Var Kent mango fruits, in four sizes, with four types of waxing. Date: 22 - 02 - 2016 (evaluation 6)

\TRATAM.	CALIBER - 6				CALIBER - 7				CALIBER - 8				CALIBER 9				TOTAL
BLOCKSx	Eo	E1	E2	E3	E0	E1	E2	E3	E0	E1	E2	E3	E0	E1	E2	E3	
I	0.448	0.448	0.448	0.448	0.736	0.448	0.368	0.394	0.460	0.576	0.333	0.499	0.544	0.384	0.400	0.310	7.244
II	0.448	0.480	0.512	0.480	0.480	0.480	0.384	0.352	0.480	0.512	0.480	0.448	0.480	0.512	0.435	0.352	7.315
III	0.480	0.416	0.480	0.442	0.448	0.512	0.480	0.500	0.544	0.378	0.500	0.416	0.448	0.544	0.429	0.448	7.465
IV	0.512	0.576	0.442	0.448	0.429	0.397	0.352	0.320	0.429	0.480	0.512	0.371	0.435	0.442	0.480	0.416	7.041
C X E	1.888	1.920	1.882	1.818	2.093	1.837	1.584	1.566	1.913	1.946	1.825	1.734	1.907	1.882	1.744	1.526	29.065
CALIBER	C6 = 7.508 0.469		Æ6	=C7 = 7.080 0.443			77	=C8 = 7.418 0.464			78	=C9 = 7.059 0.441			7P	=fG 0.454	=
WAXING	E0 = 7.801 0.488		£	=E1 = 7.585 = 0.474			71	E2 = 7.035 0.440			72	=E3 = 6.644 0.415			73		=

Acidity in Var Kent mango fruit, in four sizes, with four types of waxing. Date: 29 - 02 - 2016 (evaluation 7)

\TRATAM.	CALIBER - 6				CALIBER - 7				CALIBER - 8				CALIBER 9				TOTAL
BLOCKSx	Eo	Ei	E2	E3	E0	E1	E2	E3	E0	E1	E2	E3	E0	E1	E2	E3	
I	0.640	0.448	0.416	0.384	0.512	0.416	0.384	0.448	0.448	0.512	0.416	0.448	0.512	0.448	0.416	0.384	7.232
II	0.448	0.500	0.448	0.454	0.544	0.448	0.416	0.384	0.480	0.467	0.448	0.461	0.467	0.448	0.352	0.448	7.213
III	0.448	0.448	0.416	0.448	0.448	0.416	0.448	0.352	0.512	0.384	0.480	0.422	0.442	0.416	0.384	0.416	6.880
IV	0.435	0.416	0.544	0.416	0.416	0.384	0.416	0.384	0.416	0.352	0.480	0.384	0.429	0.390	0.448	0.416	6.726
C X E	1.971	1.812	1.824	1.702	1.920	1.664	1.664	1.568	1.856	1.715	1.824	1.715	1.850	1.702	1.600	1.664	28.051
CALIBER	C6 = 7.309 Æ6 = 0.457			C7 = 6.816 Î7= 0.426				C8 = 7.110 Jfs = 0.444				C9 = 6.816 Æ9= 0.426				Д 0.438	=
WAXING	E0 = 7.597 Υ0 = 0.475			E1 = 6.893 71 = 0.431				E2 = 6.912 72 = 0.432				Es = 6.649 73 = 0.416					

Vitamin C in mango fruit Var Kent, in four sizes, with four types of waxing. Date: 18 - 01 - 2016 (evaluation 1)

\TRATAM.	CALIBER - 6				CALIBER - 7				CALIBER - 8				CALIBER 9				TOTAL
BLOCKSx	Eo	E1	E2	E3	E0	E1	E2	E3	E0	E1	E2	E3	E0	E1	E2	E3	

																	TOTAL
I	3.50	3.00	4.00	3.90	5.50	3.50	3.10	4.00	6.00	3.50	3.30	4.20	4.50	5.50	3.50	5.70	66.70
II	4.20	3.80	3.20	6.10	4.10	4.60	5.00	4.50	3.50	4.90	4.50	5.80	3.90	4.10	3.00	5.00	70.20
III	4.50	5.00	4.90	5.00	3.20	3.90	5.00	4.90	3.90	4.50	4.80	4.50	5.10	4.80	4.00	4.10	72.10
IV	3.10	5.80	4.20	4.80	3.90	6.00	4.10	3.80	3.70	5.10	3.80	4.00	3.50	5.10	3.80	3.50	68.20
C X E	15.30	17.60	16.30	19.80	16.70	18.00	17.20	17.20	17.10	18.00	16.40	18.50	17.00	19.50	14.30	18.30	277.20
CALIBER	C6 = 69.00	Æ6 = 4.31			C7 = 69.10 £ 4.32				C8 = 70.00 X8 = 4.38				C9 = 69.10 £. = 4.32				GG = 4.33
WAXING	E0 = 66.10 Ÿ0 = 4.13				E1 = 73.10 Ï1 = 4.57				E2 = 64.20 Ÿ2 = 4.01				E3 = 73.80 Ï3 = 4.61				

Vitamin C in mango fruit Var Kent, in four sizes, with four types of waxing. Date: 25 - 01 - 2016 (evaluation 2)

\TRATAM. BLOCKSx	CALIBER - 6				CALIBER - 7				CALIBER - 8				CALIBER 9				TOTAL
	Eo	E1	E2	E3	E0	E1	E2	E3	E0	E1	E2	E3	E0	E1	E2	E3	
I	4.88	4.80	4.80	3.78	6.78	5.10	5.10	4.44	6.69	3.78	5.72	4.75	3.87	4.40	5.72	6.16	80.77
II	5.19	4.93	4.00	7.00	4.40	5.72	6.60	5.98	6.25	6.07	3.96	4.66	5.28	6.60	4.58	6.60	87.82
III	6.20	4.40	5.10	5.72	5.98	6.51	5.72	4.44	4.84	5.72	6.60	4.40	4.49	6.60	5.90	5.98	88.60
IV	5.54	5.72	4.80	4.31	6.20	4.49	5.90	5.72	5.98	5.72	4.49	6.60	6.60	5.37	6.42	4.31	88.17
C X E	21.81	19.85	18.70	20.81	23.36	21.82	23.32	20.58	23.76	21.29	20.77	20.41	20.24	22.97	22.62	23.05	345.36
CALIBER	C6 = 81.17	Æ6 = 5.07			C7 = 89.09 77 = 5.57				C8 = 86.23 78 = 5.39				C9 = 88.88 79 = 5.56				Д = 5.40
WAXING	Eo = 89.17 Æ0 = 5.57				E1 = 85.93 71 = 5.37				E2 = 85.41 72 = 5.34				E3 = 84.85 73 = 5.30				

Vitamin C in mango fruit Var Kent, in four sizes, with four types of waxing. Date: 01 - 02 - 2016 (evaluation 3).

\TRATAM. BLOCKSx	CALIBER - 6				CALIBER - 7				CALIBER - 8				CALIBER 9				TOTAL
	Eo	E1	E2	E3	E0	E1	E2	E3	E0	E1	E2	E3	E0	E1	E2	E3	
I	5.37	5.19	8.14	6.07	4.58	8.45	7.17	9.59	9.86	6.20	6.63	7.30	6.95	5.63	5.10	6.20	108.43
II	6.42	6.95	5.23	6.78	6.95	10.38	5.81	10.03	6.47	6.86	8.40	7.66	7.48	6.95	6.20	9.59	118.16
III	6.16	10.21	5.81	8.54	4.80	10.65	5.32	5.02	9.15	5.72	8.98	6.03	5.10	7.04	6.51	7.74	112.78
IV	7.57	6.10	10.38	5.54	10.21	7.08	9.06	5.50	7.83	8.14	6.47	8.45	5.50	5.19	10.25	7.57	120.84
C X E	25.52	28.45	29.56	26.93	26.54	36.56	27.36	30.14	33.31	26.92	30.48	29.44	25.03	24.81	28.06	31.10	460.21
CALIBER	C6 = 110.46 Ÿ6 = 6.90				C7 = 120.60 Æ7 = 7.54				C8 = 120.15 Ÿ8 = 7.51				C9 = 109.00 Æ9 = 6.81				Д = 7.19
WAXING	Eo = 110.40 Ÿ0 = 6.90				E1 = 116.74 Æ1 = 7.30				E2 = 115.46 Ÿ2 = 7.22				E3 = 117.61 Ï3 = 7.35				

Vitamin C in mango fruit Var Kent, in four sizes, with four types of waxing. Date: 08 - 02 - 2016 (evaluation 4)

\TRATAM. BLOCKSx	CALIBER - 6				CALIBER - 7				CALIBER - 8				CALIBER 9				TOTAL
	Eo	E1	E2	E3	E0	E1	E2	E3	E0	E1	E2	E3	E0	E1	E2	E3	
I	7.39	8.00	5.94	6.00	8.10	5.28	7.30	7.61	4.53	4.05	6.51	7.64	5.76	5.20	6.64	6.91	102.86
II	6.62	6.42	6.90	7.74	7.90	6.38	4.27	5.59	12.14	4.52	6.86	5.63	6.25	5.98	5.06	6.60	107.86
III	6.16	7.83	5.28	9.55	6.91	7.59	5.37	6.24	7.22	6.20	8.58	5.46	6.86	6.50	7.13	5.19	108.07

IV	5.76	6.20	6.85	5.32	5.46	6.06	7.58	6.34	5.72	5.32	7.30	5.01	6.94	5.24	6.25	5.63	96.98
C X E	25.93	28.45	24.97	28.61	28.37	25.31	24.52	25.78	29.61	23.09	29.25	23.74	25.81	22.92	25.08	24.33	415.77
CALIBER	C6 = 107.96 76 = 6.75				C7 = 103.98 77 = 6.50				C8 = 105.69 78 = 6.61				C9 = 98.14 79 = 6.13				Д = 6.50
WAXING	Eo = 109.72 70 = 6.86				E1 = 99.77 71 = 6.24				E2 = 103.82 7o = 6.49				E3 = 102.46 73 = 6.40				

Vitamin C in mango fruit Var Kent, in four sizes, with four types of waxing. Date: 15 - 02 - 2016 (evaluation 5)

\TRATAM.	CALIBER - 6				CALIBER - 7				CALIBER - 8				CALIBER 9				TOTAL
BLOCKSx	Eo	E1	E2	E3	E0	E1	E2	E3	E0	E1	E2	E3	E0	E1	E2	E3	
I	5.46	5.63	6.86	5.10	7.57	5.02	8.18	8.10	5.81	7.40	6.78	7.75	5.72	4.40	7.48	5.54	102.80
II	7.72	7.22	6.60	7.04	6.16	7.04	7.13	7.04	6.16	6.60	5.63	5.54	5.63	6.60	6.78	5.63	104.52
III	6.25	6.95	8.48	6.51	6.60	6.69	7.30	6.60	7.04	8.69	6.16	6.60	5.46	5.72	6.07	6.34	107.46
IV	8.07	6.69	6.51	5.98	7.92	7.80	6.95	6.94	5.19	7.48	7.39	8.50	7.48	6.16	7.22	4.93	111.21
C X E	27.50	26.49	28.45	24.63	28.25	26.55	29.56	28.68	24.20	30.17	25.96	28.39	24.29	22.88	27.55	22.44	425.99
CALIBER	C6 = 107.07 X6 = 6.69				C7 = 113.04 -I7 = 7.07				C8 = 108.72 X8 = 6.80				C9 = 97.16 Æ9 = 6.07				A' G = 6.66
WAXING	E0 = 104.24 ∎I0 = 6.52				E1 = 106.09 I1 = 6.63				E2 = 111.52 Jf2 = 6.97				E3 = 104.14 X3 = 6.51				

Vitamin C in mango fruit Var Kent, in four sizes, with four types of waxing. Date: 22 - 02 - 2016 (evaluation 6)

\TRATAM.	CALIBER - 6				CALIBER - 7				CALIBER - 8				CALIBER 9				TOTAL
BLOCKSx	Eo	E1	E2	E3	E0	E1	E2	E3	E0	E1	E2	E3	E0	E1	E2	E3	
I	6.30	7.00	6.90	7.50	6.55	5.50	7.50	8.10	6.50	6.86	8.33	6.51	6.51	5.54	7.60	7.93	111.13
II	7.80	4.61	7.60	8.13	8.30	6.51	4.60	6.72	7.55	7.12	7.30	5.98	8.50	8.30	6.90	8.44	114.36
III	8.20	7.20	7.30	6.50	6.86	7.10	6.50	7.93	6.63	8.09	5.80	7.90	7.53	7.93	7.53	7.75	116.75
IV	6.80	7.10	8.40	6.68	8.10	6.24	7.75	7.72	7.90	6.25	7.80	7.61	5.90	9.10	8.52	6.54	118.41
C X E	29.10	25.91	30.20	28.81	29.81	25.35	26.35	30.47	28.58	28.32	29.23	28.00	28.44	30.87	30.55	30.66	460.65
CALIBER	C6 = 114.02 76 = 7.13				C7 = 111.98 77 = 7.00				C8 = 114.13 78 = 7.13				C9 = 120.52 7P = 7.53				GG = 7.20
WAXING	E0 = 115.93 I0 = 7.25				E1 = 110.45 71 = 6.90				E2 = 116.33 72 = 7.27				Es = 117.94 7s = 7.37				

Vitamin C in mango fruit Var Kent, in four sizes, with four types of waxing. Date: 29 - 02 - 2016 (evaluation 7)

\TRATAM.	CALIBER - 6				CALIBER - 7				CALIBER - 8				CALIBER 9				TOTAL
BLOCK x\ ES	Eo	EI	E2	E3	Eo	EI	E2	E3	Eo	EI	E2	E3	Eo	EI	E2	E3	
I	4.44	6.51	5.98	6.78	7.30	5.19	6.69	8.45	6.43	4.84	5.63	4.49	7.92	8.10	7.22	6.42	102.39
II	5.00	5.55	7.80	8.03	8.13	8.18	6.86	6.18	8.25	7.50	8.10	4.90	6.25	6.86	7.90	6.51	112.00
III	7.90	6.90	4.50	7.80	6.51	7.60	7.75	7.13	7.61	7.90	6.87	7.45	8.45	4.94	8.10	6.88	114.29
IV	6.40	5.80	7.50	8.30	6.93	8.35	7.95	6.63	8.10	6.93	7.90	7.54	6.93	7.37	8.30	7.90	118.83
C X E	23.74	24.76	25.78	30.91	28.87	29.32	29.25	28.39	30.39	27.17	28.50	24.38	29.55	27.27	31.52	27.71	447.51

| CALIBER | C6 = 105.19 6.57 | ʹY6 = | C7 = 115.83 = 7.24 | 77 | C8 = 110.44 6.90 | â8 = | C9 = 116.05 7.25 | 7- = | GG = 6.99 |
| WAXING | Eo = 112.55 = 70.3 | ÏO | Eı = 108.52 = 6.78 | 71 | E2 = X2 = 7.19 | 115.05 | E3 = 111.39 6.96 | 73 = | |

Firmness in mango fruit Var. Kent, in four sizes, with four types of waxing. Date: 18-01-2016 (evaluation 1)

\TRATAM.	CALIBER - 6				CALIBER - 7				CALIBER - 8				CALIBER 9				TOTAL
BLOCKSx	Eo	E1	E2	E3	Eq	E1	E2	E3	Eq	E1	E2	E3	Eq	E1	E2	E3	
I	19.50	17.25	16.15	16.30	14.95	17.20	16.80	15.75	18.60	16.40	17.50	17.20	16.30	16.30	19.00	16.15	271.35
II	17.20	16.80	18.70	17.00	16.10	16.30	15.60	16.40	19.00	15.80	16.00	16.40	18.10	15.60	17.20	15.20	267.40
III	16.30	17.50	17.20	16.40	17.40	15.60	16.40	15.70	16.40	16.80	18.00	17.60	19.00	17.40	16.60	16.40	270.70
IV	18.30	15.50	15.90	15.60	18.20	18.70	18.80	16.10	17.60	18.00	18.50	15.50	17.40	16.80	18.50	17.60	277.00
C X E	71.30	67.05	67.95	65.30	66.65	67.80	67.60	63.95	71.60	67.00	70.00	66.70	70.80	66.10	71.30	65.35	1086.45
CALIBER	C6 = 271.60 16.98		76 =	C7 = 266.00 16.63			77=	C8 = 275.30 17.21			78 =	C9 = 273.55 17.10			7P =		Д = 16.98
WAXING	E0 = 280.35 = 17.52		7o	E1 = 71 = 16.75		267.95	E2 = 276.85 = 17.30			72	E3 = 261.30 16.33			73 =			

Firmness in mango fruit Var. Kent, in four sizes, with four types of waxing. Date: 25 - 01 - 2016 (evaluation 2).

\TRATAM.	CALIBER - 6				CALIBER - 7				CALIBER - 8				CALIBER 9				TOTAL
BLOCKSx	Eo	E1	E2	E3	E0	E1	E2	E3	E0	E1	E2	E3	E0	E1	E2	E3	
I	12.00	12.50	13.00	11.80	12.40	12.90	14.60	13.00	11.90	13.30	14.00	13.50	13.10	12.80	13.70	13.80	208.3
II	11.80	13.30	14.00	12.00	13.50	12.50	16.00	13.50	13.10	14.00	15.00	13.40	12.00	14.00	14.00	14.00	216.1
III	13.00	12.20	16.00	13.60	12.20	14.50	13.10	12.80	12.20	15.20	14.30	14.00	13.10	15.20	13.50	15.00	219.9
IV	12.70	12.90	15.00	11.50	11.90	13.50	14.30	11.90	12.40	16.10	13.70	12.70	13.40	13.60	15.00	13.70	214.3
C X E	49.50	50.90	58.00	48.90	50.00	53.40	58.00	51.20	49.60	58.60	57.00	53.60	51.60	55.60	56.20	56.50	858.60
CALIBER	C6 = 207.30 12.96		76 =	C7 = 212.60 13.29			77=	C8 = 218.80 13.68			X8 =	C9 = 219.90 13.74			79 =	V. =13 .42	
WAXING	E0 = 200.70 12.54		Ïo =	E1 = 71 = 13.66		218.50	E2 = 229.20 = 14.33			Æ2	E3 = 210.20 13.14			73 =			

Firmness of Var Kent mango fruit, in four sizes, with four types of waxing. Date: 01 - 02 - 2016 (evaluation 3).

\TRATAM.	CALIBER - 6				CALIBER - 7				CALIBER - 8				CALIBER 9				TOTAL
BLOCKSx	Eo	E1	E2	E3	Eq	E1	E2	E3	Eq	E1	E2	E3	Eq	E1	E2	E3	
I	11.00	12.30	12.80	12.20	11.50	13.00	13.00	12.70	11.70	12.60	13.30	13.10	10.70	11.90	12.00	12.00	195.80
II	12.20	13.00	12.10	13.00	12.60	11.00	13.40	13.00	11.90	13.00	13.00	12.90	11.20	12.00	12.80	13.00	200.10
III	11.30	11.70	13.30	12.00	12.10	12.20	12.50	12.00	13.00	11.30	12.00	12.00	13.40	12.70	13.00	12.90	197.40
IV	12.00	12.50	13.00	12.60	11.60	13.40	12.40	12.80	12.00	12.50	12.50	13.00	12.00	13.00	11.70	11.80	198.80
C X E	46.50	49.50	51.20	49.80	47.80	49.60	51.30	50.50	48.60	49.40	50.80	51.00	47.30	49.60	49.50	49.70	792.10
CALIBER	C6 = 197.00 =12 .31		.Ï6	C7 = .Ï7= 12.45		199.20	C8 = 199.80 12.49			fs =	C9 = 196.10 12.26			Æ9 =	V. =12 .38		
WAXING	Eq = 190.20		Ïo	E1 = 198.10		i1	E2 = 202.80			î2 =	E3 = 201.00			Î3 =			

	= 11.89	= 12.38	12.68	12.56	

Firmness in mango fruit Var. Kent, in four sizes, with four types of waxing. Date: 08 - 02 - 2016 (evaluation 4)

\TRATAM. BLOCKSx	CALIBER - 6				CALIBER - 7				CALIBER - 8				CALIBER 9				TOTAL
	Eo	E1	E2	E3	EQ	E1	E2	E3	EQ	E1	E2	E3	EQ	E1	E2	E3	
I	7.90	9.70	10.00	9.80	8.30	8.10	8.50	9.00	9.00	9.60	9.70	9.10	8.80	8.40	10.60	9.30	145.80
II	8.00	8.60	8.70	9.80	9.10	8.40	10.00	8.70	8.00	9.30	10.00	8.90	8.30	8.40	8.80	9.70	142.70
III	9.20	8.90	10.00	9.00	7.60	9.10	8.90	9.30	7.90	8.50	8.60	9.20	9.10	9.60	10.50	8.50	143.90
IV	7.70	8.00	9.80	9.30	8.00	8.50	9.50	9.60	8.20	8.70	9.00	8.50	7.80	8.30	8.70	8.80	138.40
C X E	32.80	35.20	38.50	37.90	33.00	34.10	36.90	36.60	33.10	36.10	37.30	35.70	34.00	34.70	38.60	36.30	570.80
CALIBER	C6 = 144.40 9.03			76 =	C7 = 140.60 8.79			77=	C8 = 142.20 8.89			78 =	C9 = 143.60 8.98			79 =	Д = 8.92
WAXING	EQ = 132.90 = 8.31			7o	E1 = 71 = 8.76			= 140.10	E2 = 151.30 9.46			72 =	Es = 146.50 9.16			7s =	

Firmness in mango fruit Var. Kent, in four sizes, with four types of waxing. Date: 15 - 02 - 2016 (evaluation 5)

\TRATAM. BLOCKSx	CALIBER - 6				CALIBER - 7				CALIBER - 8				CALIBER 9				TOTAL
	Eo	E1	E2	E3	E0	E1	E2	E3	E0	E1	E2	E3	E0	E1	E2	E3	
I	7.70	9.20	9.30	8.10	6.40	9.90	8.10	8.20	7.20	9.10	8.90	9.00	7.10	8.20	9.60	8.60	134.60
II	6.50	8.70	8.00	8.90	7.20	9.70	9.00	8.70	7.10	8.60	9.40	8.90	6.90	9.00	9.40	9.00	135.00
III	6.80	9.00	8.90	9.00	7.50	8.20	9.30	8.90	6.90	8.90	9.50	8.60	7.50	8.90	8.90	8.70	135.50
IV	7.10	9.30	9.10	8.60	6.50	8.50	8.80	9.10	6.70	9.30	8.70	8.50	7.60	8.70	8.60	8.30	133.40
C X E	28.10	36.20	35.30	34.60	27.60	36.30	35.20	34.90	27.90	35.90	36.50	35.00	29.10	34.80	36.50	34.60	538.50
CALIBER	C6 = 134.20 8.39			76 =	C7 = 134.00 8.38			77=	C8 = 135.30 8.46			78 =	C9 = 135.00 8.44			7P =	Д = 8.41
WAXING	E0 = 112.70 7.04			ío =	E1 = 71 = 8.95			= 143.20	E2 = 143.50 = 8.97			72	E3 = 139.10 8.69			73 =	

Firmness in mango fruit Var. Kent, in four sizes, with four types of waxing. Date: 22 - 01 - 2016 (evaluation 6).

\TRATAM. BLOCKSx	CALIBER - 6				CALIBER - 7				CALIBER - 8				CALIBER 9				TOTAL
	Eo	E1	E-	E3	E0	E1	E-	E3	E0	E1	E-	E3	E0	E1	E-	E3	
I	6.20	7.45	8.40	7.90	5.80	8.00	6.60	7.30	6.85	6.80	9.00	6.80	5.25	5.60	7.00	8.60	113.55
II	5.30	6.50	8.50	7.20	6.50	5.70	8.90	8.30	5.30	7.20	8.90	7.50	6.90	7.20	8.20	8.20	116.30
III	6.70	7.30	6.70	8.00	6.00	6.40	7.60	6.90	5.90	5.80	8.60	8.60	6.50	6.40	9.00	7.80	114.20
IV	6.80	6.70	7.80	8.50	5.20	7.70	8.00	7.50	6.70	6.60	7.00	8.00	6.30	8.10	7.90	8.10	116.90
C X E	25.00	27.95	31.40	31.60	23.50	27.80	31.10	30.00	24.75	26.40	33.50	30.90	24.95	27.30	32.10	32.70	460.95
CALIBER	C6 = 115.95 Æ6 = 7.25				C7 = 112.40 £7- 7.03				C8 - 115.55 X8 - 7.22				C9 - 117.05 £9 - 7.32				Д - 7.20
WAXING	E0 = 98.20 _ÏO = 6.14				E1 = 109.45 H - 6.84				E-- 128.10 2- - 8.01				E3 - 125.20 3- - 7.83				

Firmness in mango fruit Var. Kent, in four sizes, with four types of waxing. Date: 29 - 01 - 2016 (evaluation 7).

\TRATAM.	CALIBER - 6				CALIBER - 7				CALIBER - 8				CALIBER 9				TOTAL
BLOCKSX	Eo	E1	E2	E3	E0	E1	E2	E3	E0	E1	E2	E3	E0	E1	E2	E3	
I	5.50	7.60	7.60	8.00	6.10	7.30	8.50	8.90	5.90	7.10	7.30	8.30	5.70	8.90	8.80	7.30	118.80
II	6.30	6.30	8.20	8.50	5.20	7.00	7.60	7.70	6.00	6.50	7.60	7.90	5.50	8.00	7.20	7.80	113.30
III	5.70	7.00	7.60	7.50	6.30	6.90	7.90	8.50	6.50	7.00	8.00	8.70	6.30	7.90	7.40	8.10	117.30
IV	6.00	6.70	7.80	7.90	6.50	6.70	7.00	8.00	6.30	7.20	7.20	7.50	5.20	7.20	8.00	8.30	113.50
C X E	23.50	27.60	31.20	31.90	24.10	27.90	31.00	33.10	24.70	27.80	30.10	32.40	22.70	32.00	31.40	31.50	462.90
CALIBER	C6 = 114.20 76 = 7.14				C7 = 116.10 £7- 7.26				C8 - 115.00 78 - 7.19				C9 - 117.60 79 - 7.35				Д - 7.23
WAXING	E0 = 95.00 7 = 5.94				E1 = 115.30 H - 7.21				E2 - 123.70 72 - 7.73				En - 128.90 7э - 8.06				

Weight ranges and average weight of mangoes according to size (*)

CALIBER	WEIGHT RANGE (gr.)	AVERAGE
16	250-275	262
14	276-316	296
12	317-380	349
10	381-425	403
9	426-500	463
8	501-550	526
7	551-650	600
6	651-750	700
5	751-890	820
SP	>890	

Relationship of mango fruit weight and size versus dipping time for preventive fruit fly treatment

WEIGHT	CALIBERS	IMMERSION TIME
UP TO 425 grs.(1)	10,12,14 y 16	75 minutes
426-650 grams	6,7,8 ,9,	90 minutes

Note: For Kent mangoes treated at 75 minutes, for safety reasons, fruit up to 420 grams can be worked with.

Weight ranges and average weight in mangoes according to size(**)

Caliber	Weight range (gr.)
14	281-310

12	311-380
10	381-425
9	426-480
8	481-550
7	551-645
6	646-720
5	721-885

(**) REF: BIOFRUIT Company

ANNEX 40. COMPLEMENTARY FIGURES

Annex 40.1. Selection of fruit of the different sizes used in the research.

Annex 40.2. determination of fruit firmness using the Penetrometer.

Annex 40.3. Determination of brix degrees of fruit by Refractometer .

Selection of experimental units to be evaluated in the laboratory.

Annex 40.5. Samples fully identified for the determination of Acidity and Vitamin C.

Annex 40.6. Recording of data on determinations made

I want morebooks!

Buy your books fast and straightforward online - at one of world's fastest growing online book stores! Environmentally sound due to Print-on-Demand technologies.

Buy your books online at
www.morebooks.shop

Kaufen Sie Ihre Bücher schnell und unkompliziert online – auf einer der am schnellsten wachsenden Buchhandelsplattformen weltweit! Dank Print-On-Demand umwelt- und ressourcenschonend produzi ert.

Bücher schneller online kaufen
www.morebooks.shop

Printed by Books on Demand GmbH, Norderstedt / Germany